How Safe Is Safe?

How Safe Is Safe?

Radiation Controversies Explained

BARRIE LAMBERT

LONDON SYDNEY WELLINGTON

First published in Great Britain by Unwin ® Paperbacks, an imprint of Unwin Hyman Limited, in 1990.

Unwin Hyman Limited
15–17 Broadwick Street
London W1V 1FP

Allen & Unwin Australia Pty Ltd
8 Napier Street, North Sydney, NSW 2060, Australia

Allen & Unwin New Zealand Pty Ltd with the Port Nicholson Press
Compusales Building, 75 Ghuznee Street, Wellington, New Zealand

---

British Library Cataloguing in Publication Data

Lambert, Barrie
How safe is safe? : radiation controversies explained.
1. Radiation. Safety aspects
I. Title
539.7′2′0289
ISBN 0-04-440347-X

Typeset in Baskerville by Computape (Pickering) Ltd, North Yorkshire
Printed in Great Britain by Cox & Wyman Ltd, Reading

Contents

Acknowledgements

I would like to thank the following authorities for permission to reproduce the figures and tables mentioned below.

Figure 1.2 diagram reproduced by permission of the United Nations Environmental Programme; *Figure 1.3* diagram reproduced by permission of the US National Council on Radiation Protection and Measurements; *Figure 1.4, 1.8, Table 1.1* reproduced from *An Intrdouction to radiation Protection* by Alan Martin and Samuel Harbison published by Chapman & Hall, London by permission of the authors and publishers; *Figure 2.1* reproduced from: J. Rotblat, *Journal of Radiological Protection*, 8, pp. 39–46 (1988) by permission of the author and IOP Publishing Ltd; *Figure 2.2* reproduced from: S. Darby, *Health Physics* 51, pp. 269–81 (1986) by permission of the author and Pergamon Press; *Table 2.1* reproduced from: 'Carcinogenesis Following Medical Uses of Ionising Radiation' by R. H. Mole, in *Low Dose Radiation – Biological Basis of Risk Assessment* published by Taylor & Francis (1989), by permission of the author and publishers; *Table 2.5* reproduced from: 'Epidemiological Studies of Workers in the Nuclear Industry' by V. Beral *et al.*, in *radiation and Health* (Eds R. Russell-Jones and R. Southwood) published by John Wiley, (1987), by permission of the authors and publishers; *Figure 3.1* figure reproduced from publication 41 of the International Commission on Radiological Protection by permission of Pegamon Press; *Figure 3.3* reproduced from: J. S. Evans and D. W. Moeller, *Health Physics*, 56, pp. 397–413 (1989) by permission of the authors and Pergamon Press; *Figure 3.4, Table 3.9, 3.10, 3.11, 4.1, 5.3, 5.4, 5.6, 5.7, 6.1, 6.2, 6.5, 6.7, 7.2, 7.3, 8.4* reproduced by permission of the National Radiological Protection Board; *Figure 3.6* reproduced from: P. smith and R. Doll, *British Medical Journal*, 13 February 1982, pp. 449–60 by permission of the authors; *Table 3.6* reproduced from the Ionising Radiation Regulations

(Statutory Instruments 1985 No. 1333) by permission of the Controller of Her Majesty's stationery Office; *Table 3.12* reproduced from the BEIR V report by permission of the US NAS; *Figure 4.1, 4.6, 4.8, Table 4.8* reproduced by permission of the Watt Committee on Energy; *Figure 4.2* reprinted by permission of *Nature* 322, p. 19 (21 August 1986), copyright 1986 Macmillan Magazines Ltd; *Figure 4.3, 9.10, 9.11, Table 9.6* reproduced by permission of the UK Atomic Energy Authority; *Figure 4.4* reproduced from *The Transport and Deposition of Airborne Debris from the Chernobyl Nuclear Power Plant Accident with Special Emphsis on the Consequences to the United Kingdom* by F. B. Smith and M. J. Clark, by permission of the authors and the Controller of Her Majesty's Stationery Office; *Figure 5.2* reproduced from *Investigation of the Possible Increased Incidence of Cancer in W. Cubria*, Report of the Advisory Group under the chairmanship of Sir Douglas Black, by permission of the Controller of Her Majesty's Stationery Office; *Figure 5.3* reproduced from the second report of COMARE (1988) by permission of the Controller of Her Majesty's Stationery Office; *Figure 5.4* reproduced from the third report of COMARE (1989) by permission of the Controller of Her Majesty's Stationery Office; *Table 5.1* data from the International Agency for Research on Cancer, Lyon;
Table 5.2 reproduced from Sir E. E. Pochin, *British Jounral of Radiology*, 60, pp. 42–50 (1987) by permission of the author; *Table 5.5* reproduced from: 'Assessing Risks of Childhood Leukaemia in Seascale' by J. W. Stather *et al.*, in *Radiation and Health* (Eds R. Russell-Jones and R. Southwood) published by John Wiley, by permission of the authors and publishers; *Figure 6.1, 6.2* reproduced by permission of Amersham International plc; *Figure 6.3* data from UNSCEAR; *Table 6.4* reproduced from Publication 50 of the International Commission on Radiological Protection by permission of Pergamon Press; *Table 6.6* data from the US NCRP and the UK NRPB; *Table 7.1* reproduced from data in: E. W. L. Fletcher *et al.*, *British Journal of Radiology* 59, pp. 165–70 (1986) by permission of the authors and the BJR; *Figure 8.2, 8.4, 8.6, 8.8* photograph copyright BNFL; *Figure 8.7* photograph copyright CEGB; *Figure 8.9* reproduced from ACTRAM's report *Transport of Radioactive Materials for Medical and Industrial Use* (1987) by permission of the Controller of Her Majesty's Stationery Office; *Table 8.1, 8.2 reproduced from IAEA Safety Series No. 6, 1985 edition by permission of the IAEA; Table 8.3* reproduced from the 1988 Report

of UNSCEAR by permission of the United Nations; *Figure 9.3 reproduced from the 9th Annual Report of RWMAC (1988) by permission of the Controller of Her Majesty's Stationery Office; Figure 9.4, 9.12, 9.13, 9.15* reproduced from Report of Working Group II of CEC Project MARINA, *Radioactivity in North European Waters* , Fisheries Research Data Report No. 20, MAFF Directorate of Fisheries Research, Crown copyright 1989, reproduced by permission of the Controller of Her Majesty's Stationery Office; *Figure 9.8* data from BNFL Annual Reports; *Table 9.1* data from IAEA Technical Report 152; *Table 9.2* reproduced from the 9th Annual Report of RWMAC by permission of the Controller of Her Majesty's Stationery Office; *Table 9.3* data from BNFL and DoE; *Table 9.4* data from MAFF; *Table 9.5* reproduced from the MAFF Annual Aquatic Monitoring Report No. 20, the Directorate of Fisheries Research, Crown copyright 1988, by permission of the Controller of Her Majesty's Stationery Office; *Figure 10.2* reproduced from 'The National Response Plan and Radioactive Incident Monitoring (RIMNET)'. A statement of proposals. DoE (1988) by permission of the Controller of Her Majesty's Stationery Office; *Table 10.1* reproduced from 1988 Report of UNSCEAR by permission of the United Nations.

I am also indebted to Katherine Mondon for her encouragement, patience and for typing the manuscript; to Paul Mountford-Lister for computer graphics; and to William Shuster and Keith Gibson for photography.

Preface

This book has been written to try to allay fears, to put things into perspective and to throw some light on radiation risks – how they are created and what the government does to protect people from them.

Each chapter examines a different aspect of radiation in our lives today. Some are more controversial than others. This is not meant to be a textbook of radiation hazards – there are enough of these already, but more a selection of those issues which, in the author's experience, are the most important or misunderstood.

Although radiation risks are better studied than most, it is clear that the public perception of these risks is very different from official estimates. There is no easy solution to this dichotomy but the establishment should be aware of public concern and err always on the side of caution. In addition, the public are often anxious about inconsistencies in official pronouncements; this loss of credibility, which does little to inspire confidence, is explored in detail particularly in Chapter 4. Reliance on expert opinion in respect of a hazard which cannot be appreciated by any of the senses produces a feeling of vulnerability, exacerbated by evidence of lack of governmental planning. An example of this was the UK government's reaction to the Chernobyl incident, when, for a time, chaos reigned. It was astonishing, and indeed rather worrying, that after thirty years of world-wide nuclear power, we seemed unprepared for such an incident.

The recent announcement by the government that no more nuclear power plants are to be built (at least until 1994) makes very little difference to the sources of radiation risk in the UK. The assured continuance of the reprocessing of nuclear fuel, if for no other reason than the huge investment in the Sellafield site, will ensure some risk both to workers and the public for the forseeable future. We still have the problems of decommissioning of power plants and storage of waste to solve. Lastly, whatever we decide to do, nuclear power continues to burgeon just across the Channel where the French, for instance, have nearly seventy nuclear power reactors and two reprocessing plants. All these are near enough to cause much greater problems in the UK than Chernobyl, in the event of a serious accident.

Radiobiology is a complex subject and although some of the concepts are explained in the text, others are covered in the glossary. In addition, the histories and roles of some of the important nuclear institutions and committees are explained and commented upon there. Hopefully, the reader will find the glossary a useful source of background information *and* comment.

In order to make the text clearer and more readable no references are given. However, a further reading list is supplied which, although not exhaustive, refers to most of the source material.

How Safe Is Safe?

1 Radiation and Its Effects

This book is concerned with the highly emotive and controversial subject of risk in the use of radiation. It will cover the sub-sciences of radiation biology, radiation protection and health physics, and touch on radioecology and epidemiology – not an easy task in one small volume. There will need to be some glossing-over of fundamental aspects in order to allot more space to contentious issues. However, so much of the science of radiation is misunderstood and there are so many myths to dispel that a chapter on basics must be included. This chapter is therefore essential reading and seeks to answer succinctly two questions: 'What is radiation?' and 'How does radiation affect people?'

The bulk of the information here will be presented as if it consists of incontrovertible facts, whereas in reality in science there are few such beasts. Where there are several theories these have been mentioned, but space will not allow a complete discussion of the pros and cons of each unless they are central to a controversy.

What Is Radiation?

Radiation that might affect us comes either as waves (so-called electromagnetic radiation) or as streams of particles (particulate radiation). This book is concerned with ionizing radiation, a type that has a damaging effect on biological tissue. The property of being capable of ionizing, that is splitting off electrons, is a characteristic of both wave and particulate radiations. In other respects the wave radiation associated with radioactivity is merely part of a spectrum or continuum of radiation that is all around us and is totally familiar. This continuum varies in frequency from radio waves through visible light to ultraviolet and X-rays and,

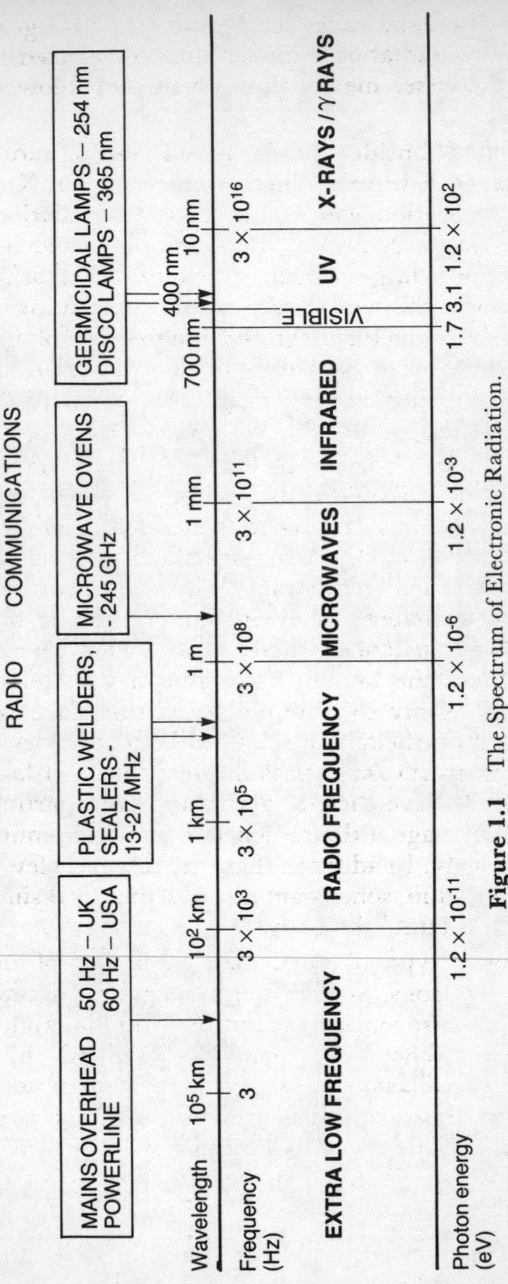

Figure 1.1 The Spectrum of Electronic Radiation.

eventually, gamma rays (see Figure 1.1). One general problem with ionizing radiation is that it goes undetected by our senses, which, of course, means that we do not know we are being irradiated.

Let us first consider those types of ionizing radiation that are electromagnetic waves. Whether created in an X-ray apparatus or by disintegration of an atom, this sort of radiation is essentially the same. In an X-ray set, radiation is created by accelerating electrons into a target (much as in a TV tube), and the energy of the resulting radiation depends on the accelerating voltage. When a radioactive atom disintegrates, processes involving the orbiting electrons may occur that result in gamma radiation of characteristic energy being emitted in packets called photons. Electromagnetic radiation of this type is extremely penetrating, so that shielding against it has to be made of lead or concrete.

Particulate radiation consists of alpha (α) and beta (β) particles, and neutrons. These are generated or emitted by atomic disintegration or nuclear reaction. Alpha and beta particles, being charged, lose their energy rapidly by collision or interaction with atoms in the material they encounter and are therefore easily shielded. For instance, whereas X-rays are capable of passing right through the human body and may therefore be used to produce an X-ray shadow picture, a substance emitting alpha particles may be held in the hand without harm because the alpha particles it emits cannot pass through the outer layer of the skin (Figure 1.2). Nevertheless, alpha and beta particles can cause biological damage if the radioactive materials emitting them get inside the body. In addition there are relatively few radioisotopes that do not emit some gamma rays during disintegration and these can penetrate the body.

The other type of particulate radiation of interest is the neutron. Neutrons, as their name suggests, are uncharged particles. They cause ionizations indirectly in material, but are fairly penetrating. They are generally emitted by an atomic rearrangement known as fission in which an atom splits roughly in half (creating two other elements) with emission of neutrons and energy. The fission of uranium is the basic process in a nuclear reactor. Neutrons are best shielded by materials containing a relatively large amount of hydrogen because neutrons lose energy by interaction with hydrogen atoms. Neutrons really only

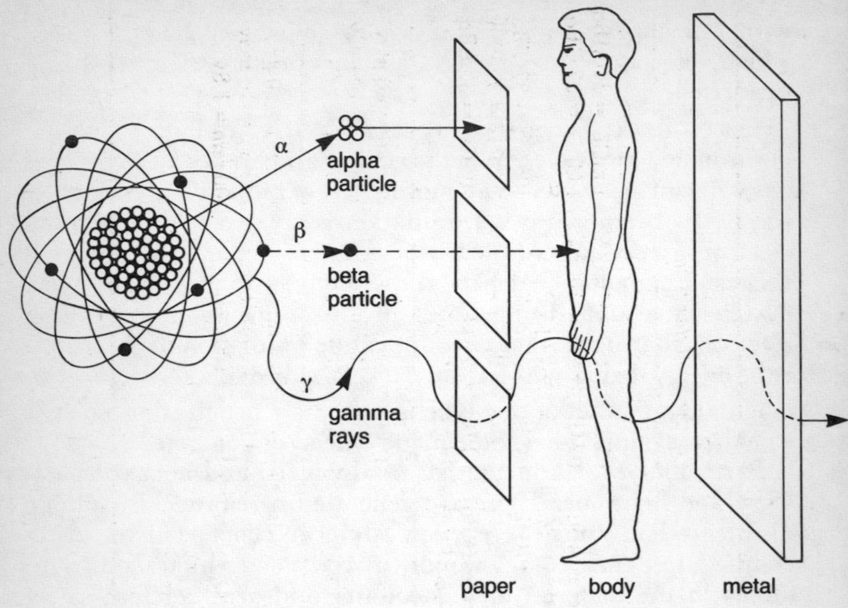

Figure 1.2 Three Types of Radiation and Their Powers of Penetration.

present a hazard to nuclear power workers because it is only in the vicinity of reactors that they are found.

Thus it can be seen that ionizing radiation is emitted by man-made machines (X-rays) or from man-made radioisotopes, but it also comes from naturally occurring radioactive materials (see Chapter 6). Whatever the source, the radiation can, in the appropriate location, cause damage to biological cells and structures.

A Brief Word About Units

The units associated with radioactivity and radiation are often misunderstood, particularly as old units have been replaced by a new set of SI units within the last ten years (Table 1.1). The unit of radioactivity currently used is the becquerel (Bq). This quantifies the rate of disintegration or decay of a radioisotope (1 Bq being 1 radioactive disintegration per second). The fundamental

Table 1.1 The Relationship of Old and New Units

Radiological quantity	*Old unit*	*SI unit*	*Relationship between units*	
Activity of a radioactive material	The curie 1 Ci = 3.7×10^{10} dis/s	The becquerel 1 Bq = 1 dis/s 10^3 Bq = 1 kilobecquerel (kBq) 10^6 Bq = 1 megabecquerel (MBq) 10^9 Bq = 1 gigabecquerel (GBq) 10^{12} Bq = 1 terabecquerel (TBq) 10^{15} Bq = 1 petabecquerel (PBq) 10^{18} Bq = 1 exabecquerel (EBq)	1 Bq = 2.7×10^{-11} Ci 1 kBq = 2.7×10^{-8} Ci 1 MBq = 2.7×10^{-5} Ci = 27 μCi 1 GBq = 27 mCi 1 TBq = 27 Ci 1 PBq = 27 kCi 1 EBq = 27 MCi	1 μCi = 37 kBq 1 mCi = 37 MBq 1 Ci = 37 GBq 10^3 Ci = 37 TBq 10^6 Ci = 37 PBq 10^9 Ci = 37 EBq
Absorbed dose	The rad 1 rad = 0.01 J/kg	The gray 1 Gy = 1 J/kg 1 Gy = 10^3 mGy = 10^6 μGy	1 μGy = 0.1 mrad 1 mGy = 100 mrad 1 Gy = 100 rad	1 mrad = 10 μGy 1 rad = 10 mGy 100 rad = 1 Gy
Dose equivalent	The rem 1 rem = 1 rad × Q Q is the quality factor	The sievert 1 Sv = 1 Gy × Q × N 1 Sv = 10^3 mSv = 10^6 μSv N is the product of all other modifying factors (currently taken as 1 by ICRP)	1 μSv = 0.1 mrem 1 mSv = 100 mrem 1 Sv = 100 rem	1 mrem = 10 μSv 1 rem = 10 mSv 100 rem = 1 Sv

law of radioactive decay states that the rate of decay or disintegration varies with the number of unchanged atoms present. Thus with time the number of becquerels, or the amount of 'activity' in a sample of radioactive material decreases. The time taken for the number of becquerels to halve is characteristic of a particular radioisotope and is called the half life (or half time). Half lives vary from a fraction of a second to billions of years, but often a radioisotope decays through a series of 'daughters' before reaching a stable non-radioactive configuration.

The becquerel is a relatively recent innovation; the traditional unit was the curie, which is the amount of radioactivity associated with 1 gram of radium. This was equivalent to 3.7×10^{10} disintegrations per second or 3.7×10^{10} Bq. The curie is a very large unit and diminutives like millicurie (a thousandth, 10^{-3}) and microcurie (a millionth, 10^{-6}) were therefore commonly used. The becquerel by contrast, is unfortunately a very small unit and large prefixes have often to be used, e.g. terabecquerel (10^{12}) and petabecquerel (10^{15}). In comparison to the curie, however, the becquerel is a very easy unit to understand and use.

Interactions of Radiation with Biological Tissue

When radiation passes into any material some energy is lost by the incident radiations and is imparted or deposited in the material. If this material is biological tissue, the measure of this deposited energy per unit mass is called the absorbed dose. The currently used unit of dose is the gray (Gy), which is 1 joule absorbed per kilogram of tissue.

It will be clear from what has been said before that certain types of radiation (e.g. alpha particles) will not travel far in absorbing material. They lose their energy quickly over a very short track or path length (about 4 cm in air or 40 μm in tissue). They have what is known as a high linear energy transfer (high LET). Because of the high density of ionizing events characteristic of radiation of this type (mostly alpha particles and neutrons), they are more damaging, dose-for-dose, than gamma or X-rays. In order to be able to sum doses from radiations of different LETs, each radiation is assigned what is called a quality factor (Q). For alpha particles, Q is 20; for gamma rays, X-rays and beta particles, Q is 1. When multiplied by this (normalizing) factor Q, absorbed dose becomes dose equivalent expressed in sieverts (Sv). Again this unit is large and milli- and micro-sieverts are commonly used.

The relationship between the radiation units is complex; for instance, there is no simple direct relationship between a concentration of a radioisotope, say, in food (in Bq/kg) and the dose delivered (in Sv) after intake, although with some assumptions about metabolism etc., it is possible to calculate the latter from the former.

Instruments for Measuring Radiation

In a book that is essentially about the risks of radiation exposure, a section on instruments may seem out of place. However, one aim of this book is to provide information and sort out misapprehensions, and there are many of these associated with the measurement of radiation. This is not an in-depth discussion of the design and use of instruments but rather a cursory guide that may correct some misconceptions.

Instruments for radiation detection and measurement do no more than register the number of ionizing events that have occurred in their sensitive volume in a certain period of time. The recording device, be it a meter or a digital (LED) display, may well indicate counts per second, total counts or a dose rate, but this is rather more a result of software than instrument design.

Instruments for radiation measurement fall into roughly two groups divided by the basic principles used in the design of the counting head or probe. One group uses gas ionization and the other has solid state detectors. By far the most robust and commonly used are gas-filled tube detectors: these include the well-known Geiger counter and proportional and ionization chamber detectors used for special purposes. All these instruments rely on detection of the very small electrical pulses generated by the ionization of a gas (by the incident radiation) and the collection of the resulting ions by two charged plates. The instrument will, of course, only work if the incident radiation does in fact ionize the counting gas. So radiation must be able first to enter the counter through the receiving 'window'. Most Geiger counter windows are made of glass that is sufficiently thick to exclude totally low-energy beta particles and allow through only some of the energy of other more energetic beta particles. This sort of counter is therefore completely inappropriate for counting, for instance, tritium (H-3),carbon-14 and sulphur-35, which decay with the emission of weak beta particles. However, counters are available with thin, fragile windows that can be used for this purpose. Clearly, there are fewer counters that will measure alpha particle emission. Alpha particles tend to be emitted at nearly all the same or similar energies and consequently travel about the same distance in air – this distance, it must be remembered, is only about 4 cm.

When alpha and beta particles get into the counting gas, they are efficiently detected, but the same cannot be said for X-rays and gamma rays, especially those of high energy. These radiations are extremely penetrating and tend not to leave very much energy behind in a small gas counting tube. This problem may be overcome by increasing the volume of the counting tube, but unfortunately this increased sensitivity has to be traded off against a higher sensitivity to 'background', which is the sum of all the 'events' recorded when the radioactive sample is not present. Background, which may come from a number of sources as diverse as natural radioactive decay and electronic 'noise' in the counter, often limits the sensitivity of radiation measurements.

Solid state detectors are more useful counting instruments for X-ray or gamma ray measurement. In these, typically, the incident radiation causes a small flash of light to be emitted from a solid 'phosphor' (for example, a large crystal of sodium iodide). These flashes are turned into electrical pulses by a photomultiplier (PM) tube and counted electronically. These instruments (scintillation counters) are up to twenty times as sensitive as Geiger tube type counters for X-rays and gamma rays but are not as robust because of the fragility of the PM tube.

Some solid state detectors depend on changes of conductivity. These have the highest resolution for gamma radiation (see Figure 1.3) but are not very efficient (typically about 10 per cent or less than the sensitivity of a scintillation detector). In addition, such semiconductor detectors, notably those using germanium, have to be operated at the temperature of liquid nitrogen (about −196°C). Their application is therefore limited to gamma spectrometric analysis in the laboratory.

For robust, general purpose detection of radiation in field conditions, a Geiger-tubed counter is adequate. With rather more

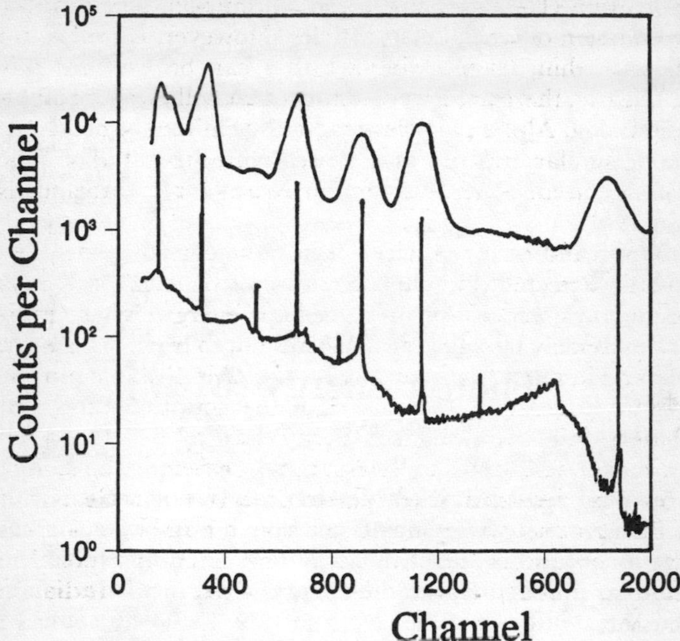

Figure 1.3 Gamma-ray spectra of a 5-ml mixed-radionuclide-solution source taken with the source within a 5-inch NaI(Tl) well crystal (upper curve) and at the face of a 60-cm^3 Ge(Li) detector (lower curve). The counting time in each case was 2000 s. (From measurements made at the National Bureau of Standards.)

sophisticated software and some assumptions about the energy of the incident radiation, such an instrument may be calibrated in dose or dose rate (i.e. say, μGy/hour). It must be remembered that this dose rate is the external dose rate from incident radiation only.

Accurate calibration is crucial. The Ionizing Radiations Regulations (1985) in the UK (see Chapter 3) require monitors used in the work place to be calibrated, but it is a good procedure for *all* instruments. It is particularly appropriate if the energy response of the instrument is not constant, that is, not 'flat' (see Figure 1.4) or if it is to be used with odd source geometry. For instance, food

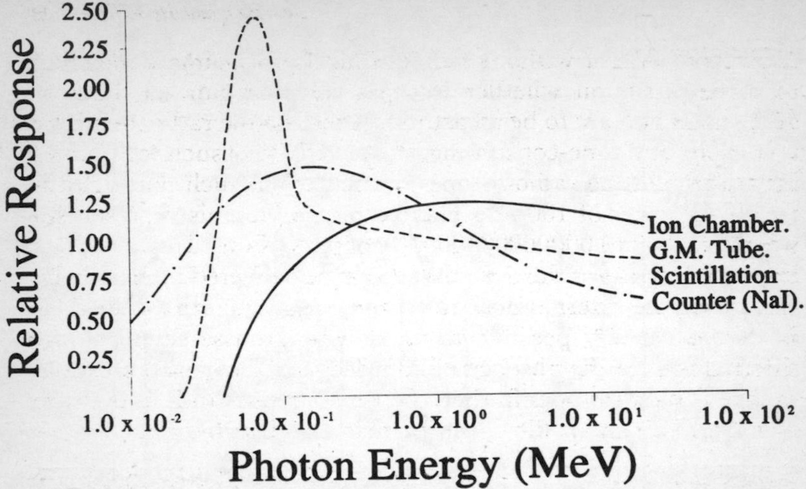

Figure 1.4 Energy Response Curves for Three Types of Radiation Detector.

monitors need appropriate (dispersed) calibration sources if they are to give accurate assessments in, say, Bq/kg. In this respect scintillation counters have an advantage in that as the pulse amplitude is proportional to the energy of the incident radiation photon, the counter may be electronically 'gated' to count only the radiation coming from one isotope. This is an extremely useful facility because it also reduces the general background count and therefore increases the signal-to-noise ratio. This can be extended by using a doubly 'gated' counter (a double-channel analyser) in which one channel is set for high-energy background and the other is used for the isotope of interest. By measuring ratios of these channels in different locations or with different sources, a more reliable estimate of background may be made.

One of the more important features of monitoring for radiation is the need for measurements to be statistically significant. This may involve a lengthy counting time. For instance, because of the randomness of radioactive decay, few modest counters are able to give a reliable estimate of the natural external background dose rate (about 0.04 μGy/hour) with less than a 5-minute integration period.

The measurement of the external gamma dose is only one part of the assessment of risk. Another is the measurement of the concentration of radioactive materials in diet and air. These measurements can be carried out extremely accurately in the laboratory with or without radiochemical processing, depending to some extent on whether isotopes emitting gamma, beta or alpha radiation are to be measured. Although the radiochemistry is complex and time-consuming, the sensitivity of such techniques for alpha-emitting radioisotopes particularly is such that quantities of the order of 100 μBq may be measured (this represents a weight of about 0.00000000000004 g of plutonium). However, it is impossible to assess the dose that might be delivered internally by measuring the external dose rate from, say, a quantity of food. It is, of course, only possible to make a rough assessment of the internal dose in the absence of knowledge of the gastrointestinal uptake of radioisotopes in diet, their metabolism and, in the case of material on an air filter, the particle size distribution.

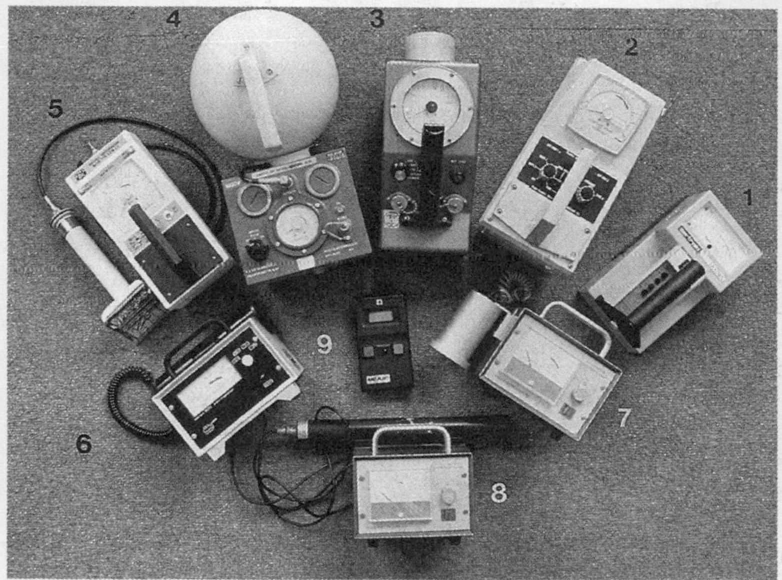

Figure 1.5 Some Radiation Monitors.

1 α/β surface contamination monitor (gas proportional counter).
2 Ionization chamber dosemeter.
3 Scintillation dosemeter.
4 Neutron monitor.
5 α/β dual phosphor surface contamination monitor.
6 Scintillation probe contamination monitor.
7 Thin window geiger contamination monitor.
8 Large volume geiger monitor.
9 'Pocket' geiger dosemeter.

Figure 1.6 An Environmental Gamma Counter.

In summary, it can be seen that instruments fall into two groups that differ in sensitivity. Choice of suitable monitoring instruments (some that are available are shown in Figures 1.5 and 1.6) is of paramount importance and there should be an awareness of their limitations for measuring low levels of radioactivity or dose. Laboratory radiochemical methods and equipment for measuring low levels of radioisotopes have developed in recent years and techniques are now available for coping with the greatest demands for sensitivity.

Personnel Dosimetry

One of the requirements of the protection regulations for radiation workers is an assessment of the dose received. None of the counters and dosemeters referred to above can do this, as what is required is an integrating (summing) device that may be carried around by the worker. Because of these requirements, the techniques of personal monitoring have advanced quickly and there are now a number of devices that can be linked to computerized record-keeping facilities on a large scale. Most of the more sensitive devices suffer from being retrospective: the dose received is only revealed when the device is read, possibly a week or two later. Clearly, tailoring the monitoring period to the risk involved in a particular job is of some importance for control of exposure. The main methods used for monitoring the doses received by workers are listed below (see Figure 1.7).

Film badges

Traditionally, film badges are the most common dosemeters and consist of a piece of photographic film in a special holder. Measurement depends on the fact that ionizing radiation blackens film in proportion to the dose received. The film has two emulsions to measure a wide range of doses, and, when developed, can be kept as a permanent record. Filters made of different materials in various parts of the holder (see Figure 1.8) allow some estimation of the quality of the incident radiation. This type of dosemeter does, however, suffer from the disadvantage of being a retrospective means of measuring radiation doses and is fairly bulky; it could not easily be used for estimating hand or eye doses.

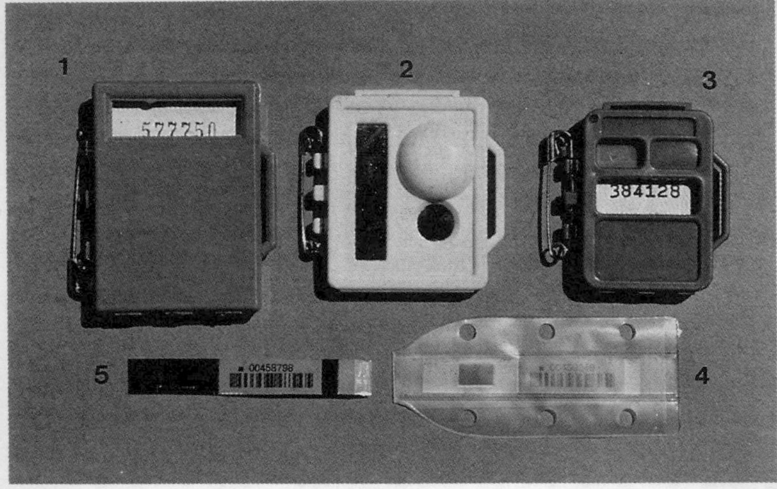

Figure 1.7 Personal Radiation Monitors.
1 Neutron film badge.
2 TLD skin and body dosemeter.
3 Film badge.
4 TLD extremity dosemeter (in finger stall).
5 TLD extremity dosemeter.

Moreover, it can only record the radiation that happens to impinge on it and there have been suggestions that it may underestimate doses by up to a factor of two, depending to some extent on where it is worn.

Thermoluminescent dosemeters (TLDs)

The use of thermoluminescent materials as dosemeters has increased in recent years, mainly because TLD readings are easily linked to computerized dose-recording systems. TLDs utilize a powder, most commonly lithium fluoride, in a small sachet. The relatively small size of this sachet is another advantage. It enables TLDs, for example, to be used as finger dosemeters or, taped to the forehead, as a means of estimating eye dose. 'Reading' of the device, which involves heating the chips to release the 'trapped' energy, destroys the information, so TLDs can only be read once. The thermoluminescent material used has a fairly good energy response but, of course, no information on the quality of the incident radiation may be obtained. Both the photographic film

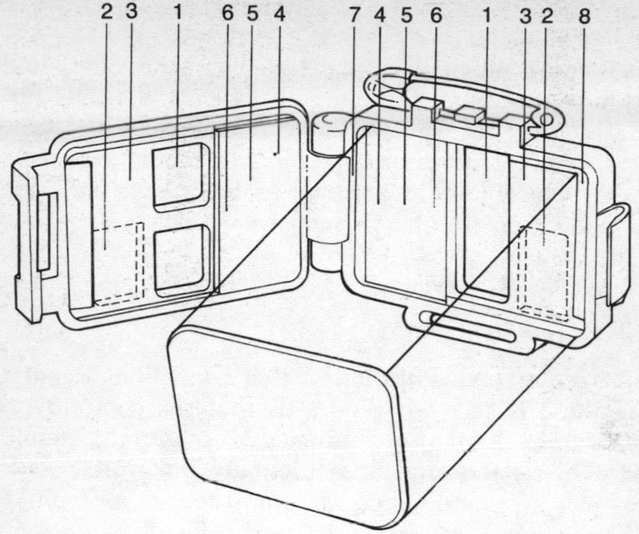

Figure 1.8 Film Badge Holder with Filters.
Filter Types
1 Window
2 50 mg/cm² plastics
3 300 mg/cm² plastics
4 0.004″ Dural
5 0.028″ Cd 0.012″ Pb
6 0.028″ Sn 0.012″ Pb
7 0.012″ Pb edge shielding
8 0.4 gm of indium

badge and the TLD have a lower limit of sensitivity of about 50 μSv.

Quartz fibre electrometers (QFEs)

The need for a dosemeter that gives a constant visual record of accumulated dose is partly answered by the quartz fibre electrometer (QFE). This consists of a quartz fibre in a small ion chamber about the size of a large pen. The decrease in the deflection of the fibre is calibrated in dose and can be used to indicate when a dose limit has been reached. The sensitivity of these devices is, however, seriously limited, i.e. a lower dose limit of about 1 mSv is usual.

A much more efficient device based on the photodiode is currently being developed. This will not only digitally display accumulated dose, but will also have an alarm and be capable of interfacing with all types of dose record-keeping systems. The dosemeter will be about the same size as the existing film badge; it remains to be seen whether it has the necessary sensitivity to replace the film.

Personal air samplers (PASs)

One of the requirements of the Ionizing Radiations Regulations (1985) in the UK (see Chapter 3) is that the committed dose equivalent from intake of radioactive materials should be recorded. This is the total dose (actually integrated over fifty years) to which an individual is committed as a result of the intake. This requirement presented great difficulties to operators, particularly those that had workers exposed to long-lived actinides, e.g. plutonium. In the past, control of environmental contamination with passive or static air samplers linked with irregular measurements of excretion and occasional whole body measurements were considered sufficient. For some radionuclides, e.g. tritium, biological (excretion) monitoring is adequate because metabolic models are fairly well established. However, for others, e.g. plutonium, in-vivo (that is, counting externally the gamma emissions from *within* the body) and urine analysis techniques are considered insufficiently sensitive to detect the annual change in organ (e.g. the lung) deposition corresponding to a fraction of the dose limits. Because it is recognized that the main problem is inhalation, most operators, on the advice of the Health and Safety Executive (HSE), chose to use personal air samplers (PASs).

These measure the material deposited on an air filter through which air from near the face has been sucked. They are used to indicate intakes related to the derived limit; the annual limit of intake (ALI) (see Chapter 3). Capital investment in these devices has been large – for instance, British Nuclear Fuels plc (BNFL) initially spent £2.2 million – but their use in relation to intake estimation is the subject of considerable controversy. They seem to show great variability even in a constant air concentration.

Even as indicators for further investigation by, for example, radiochemical analysis of nasal mucus or faecal monitoring, they are reported to be unreliable. In fact, published results show that intake estimates based on PAS can be a factor of twenty-five above or below other estimates.

The general conclusion seems to be that PAS results are not reliable enough to make acceptable assessments of intake. In these circumstances, it is difficult to understand how operators using them can effectively comply with the legal requirements of the HSE or how they are making a satisfactory estimate of the risk to their workers.

The Biological Effects of Radiation

Although this chapter is intended to be a brief review of radiation and the consequences of exposure to it, it is worth dwelling on its biological effects because an understanding of the underlying mechanisms may put some of the controversies into perspective.

As already described, ionizing radiation transfers energy from a source to an absorber, which may be human tissue. This phenomenon is somewhat similar to the transfer of heat and light from the sun to the earth but of course we are not able to sense ionizing radiation absorption. The deposition of energy in the body from a level of gamma radiation that would almost certainly result in death would raise the body temperature by less than one-thousandth of a degree centigrade. However, at any level of exposure *some* damage will occur in the body tissues and it is the quantification of this damage that is at the basis of many current controversies. Unfortunately, the whole picture is nowhere near clear; many of the observational data are incomplete and subject to statistical interpretation. Thus many of the predicted effects are the result of educated speculation based on models. Nevertheless, there are some undisputed data that give a clue to possible mechanisms of, for example, cancer induction and the quantification of risk.

First, we know that radiation exposure can cause the death of cells. This has been seen in studies in cell culture (*in vitro*) systems, i.e. in 'test tubes', and it has also been observed in humans. If too many cells are killed, the body will be unable to replace them and death may well result. This was observed in Japan after the atomic bombings in 1945, at Chernobyl in 1986 and in various other accidents. At high and low doses, radiation may also trigger off a chain of events that lead to effects seen many years later. The most common and significant of these 'late' effects of radiation is cancer. It is worth emphasizing at this point that the induction of cancer is governed by the laws of probability and that therefore any exposure increases risk without giving a certainty of effect. In addition, although a practical threshold may seem to exist (and this cannot really be totally excluded) there is ample evidence from theory that no radiation dose is totally without risk. Another point worth making is that cancer induction by radiation is a rare event and that of course radiation is by no means the only carcinogen in the environment.

Mechanisms of Radiobiology

After study of the effects of radiation for the last 30–40 years, certain basic mechanisms have been established. We are still probably some way from understanding the process completely but mechanistic theories now can be used to explain the observations.

The sequence of biological events begins with the ionization of molecules in tissue. As the organs of the body are composed of 65–70 per cent water, it is likely that initial ionizations will occur in water molecules. Once this has occurred, in the next 10^{-9}–10^{-3} seconds the electron and the positively charged ion may react with other water molecules to produce particularly reactive chemical species called free radicals. These diffuse rapidly through tissues but rarely have a lifespan longer than 10^{-5} seconds before they capture an electron. The most important reactions that aqueous free radicals undergo are with biochemical molecules and with oxygen. The results of these reactions may be subtle damage to molecules or the production of peroxides, which can cause even more damage.

The overall result of these events, which may be all over in less than a millisecond, could be changes in cells that are immediately lethal to them or subtle changes in the gene code. Interest has switched over the last few years towards changes in DNA that cause oncogenes to be activated or switched on. Oncogenes are small pieces of gene code in the DNA that have been seen to be activated in cells that have been transformed, that is, cells in which the first steps towards malignancy have occurred. The actual role of oncogenes in all cells is not positively known although some are known to code for, or control, the production of growth factors. This is significant because one of the characteristics of cancer is a local loss of control of cell division (growth).

The process described above is the 'indirect' theory of radiation effects. There are two pieces of evidence that support it. The first is the relatively high radiosensitivity of biological systems irradiated when wet rather than dry. The second is the relatively high radiosensitivity of fully oxygenated tissue; oxygen is known to promote the effect of free radicals.

Radiation (particularly high LET radiation) can also cause damage by the actual physical breaking of chemical bonds. Nevertheless, the end result is much the same as for the indirect effect. Of course, this sort of damage is occurring millions of times a day, not just from radiation exposure but also from the other toxins in our environment. For instance, natural background radiation causes more than 215-million ionizing events per hour every day in our bodies. The fact that radiogenic cancer is such a rare event is an accolade to our immuno-surveillance system and repair mechanisms. Ironically, one theory of cancer induction suggests that it occurs because of faulty repair of damage.

The most sensitive tissues are those in which there are a large number of dividing cells, e.g. the bone marrow and the lining of the gastrointestinal tract. This is strictly true for cell death and more or less true for cancer induction. Radiation is different from other carcinogens in this respect because it can induce cancers in most tissues rather than just a target organ. However, the radiosensitivity of different organs and tissues to cancer induction is markedly different and it is not always obvious why this should be so. A general pattern of radiation-induced cancer is that the cancers that occur are characteristic of the species but occur in excess, although at about the same time in life as normally expected. Thus another feature of radiogenic cancer is the latent or lag period that elapses before the overt expression of the malignancy. This has been observed to be from less than two years (for leukaemia) to in excess of forty years. Little is known about what happens during this latent period except that it is thought that radiation-induced damage may be only one step in a complex multi-link chain of causation. It is now thought that most cancers arise from a single cell and that following the initial transformation of that cell two other stages, promotion and progression, have to take place before a cancer is produced. Radiation can be either the initiator or the promoter, or both, and as there is no way of recognizing a radiation-induced cancer from one occurring from some other cause, the role of radiation in the process may remain difficult to unravel.

Acute Effects of Radiation

Some discussion of acute effects is necessary, but only for completeness, because such effects are only seen after very high doses of radiation. In the cases where either whole-body or part-body irradiation produces cell death beyond the system's repair capability, death may result in a matter of days or weeks. Deaths by this means (often called radiation sickness) were seen in 29 of the 31 people who died immediately after the Chernobyl disaster. These acute effects (often called non-stochastic effects, because their severity depends on the dose received) are only manifest after a dose threshold has been exceeded. The thresholds for most non-stochastic effects are in the range of 1–2 Sv which means that they are only usually a consequence of a serious radiation accident. However, the killing of cells caused by large doses of radiation is also utilized in radiotherapy. Tumours generally contain many dividing cells, which are consequently radiosensitive, and may be killed by treatment regimes using collimated beams of X-rays or gamma rays. The efficacy of the treatment depends on a number of factors, one of which is the dose that can be tolerated by surrounding normal tissue. The treatment often has to be given in a number of fractions, and from different angles to spare this normal tissue. Even so, experience has shown that treatment using radiation not only 'cures' the primary tumour, but also induces cancers in other tissues peripherally irradiated (see Chapter 2).

Types of Genetic Damage

Genetic information is transferred from generation to generation in the form of gene codes in the DNA of a cell's nucleus. Mutations or rearrangements in these codes, which give rise to changes in the cell's progeny that may or may not be viable, occur spontaneously or can be induced by external influences. One of these influences is radiation, although its effect on the human genome can only be speculated on because so far there have been no recognized heritable defects in humans exposed to radiation. However, it has been fairly easy to demonstrate and quantify genetic damage in lower animals and insects after irradiation.

There are three main types of gene mutation: dominant, recessive and X-linked. Each individual receives a set of genes from each parent, so that a dominant gene will be expressed if received from only one parent. However, a recessive mutation will not be expressed unless it is received from both parents. Females have two X-chromosomes and males one X and one Y. Thus, an X-linked recessive mutation can readily be expressed in a male but in a female it would have to be present in both X chromosomes. The expression of genetic diseases is complicated by the fact that additional factors such as environment can play a part. These are known as multifactorial diseases.

Most serious mutations that are inherited and are viable lead to gross mental and/or physical abnormalities. Sufferers from these conditions seldom reproduce to pass on the mutation and so the mutant load is only maintained by environmental or other influences. Dominant mutations show up in the first generation after they have arisen but do not necessarily prevent child-bearing. Recessive mutations may only occur several generations later and it is therefore important to assess radiation effects over a number of generations to see if they have shown complete expression.

Genetic Effects of Radiation

So far no genetic or heritable effects have been seen in human populations exposed to radiation. Even in studies of 30,000 children born to parents irradiated at Hiroshima and Nagasaki in Japan, there have been no differences in stillbirths, congenital malformations, neonatal deaths, malformations, birth weight or sex ratio compared to 45,000 controls. The range of doses in this study was 0.08–2 Gy (average 0.15 Gy). Although no statistically significant effects were observed, a lower limit for the dose required to double the spontaneous mutation rate, the doubling dose, was 'estimated' as 1.4–1.8 Sv.

However, because there are no direct human data, animal studies have been used to deduce the risk of radiation-induced hereditary disease. Two methods have been used:

(a) The direct method assumes that the frequency of induction of certain inherited diseases is the same per unit dose in man as in animals. Then, the total radiation-induced mutation rate is deduced as the ratio of the natural incidence of these specific disorders in man to that for all inherited disease.
(b) The indirect method uses the concept of a doubling dose, which has been mentioned above. In the mouse, the doubling dose for many different inherited disorders is about 0.3 Gy for a single dose (and 1 Gy for chronic irradiation). It has been assumed that a similar dose will double the spontaneous mutation rate in man. If it is further assumed that high dose rate exposure is about three times more effective than low dose rate, the figure obtained is similar to that estimated for the A-bomb survivors.

Risk Rates for Genetic Effects

The risk rate for genetic effects in the first two generations after exposure of either parent was adjudged by the International Commission on Radiological Protection (ICRP) in 1977 to be 1 in 100 per Gy (or 2 in 100 for all generations). This is based on a chronic exposure doubling dose of 1 Gy. In the most recent (1988) report of the United Nations Scientific Committee on the Effects of Atomic Radiation (UNSCEAR) the risk of genetic damage for all generations was estimated to be 1.2 per 100 per Gy; this includes an allowance for recessive disease, but not congenital abnormalities or multifactorials.

Clearly, for radiation-induced genetic effects to occur, the reproductive organs have to be irradiated before or during reproductive life. Therefore for application to a population, the UK's National Radiological Protection Board (NRPB) assumes a genetically significant risk coefficient of 0.8 per 100 per Gy for all generations for low dose rates.

Generally speaking, in the absence of any positive evidence of radiation-induced genetic damage in humans, the regulatory committees have assumed that the application of protection limits for somatic effects will reduce genetic effects to an acceptable level. In the absence of concrete evidence, it is as well to be cautious and it is well worth quoting the adage: 'No evidence of effect is not the same as evidence of no effect'.

Irradiation In-utero

There is now considerable evidence that some effects occur with greater frequency or are more severe for the developing embryo and foetus than for other groups in the population. The risks encompass prenatal and neonatal death, severe mental retardation and enhanced sensitivity to cancer induction, but human evidence is available only for the latter two.

Mental retardation following radiation exposure in-utero is currently receiving some attention. The period of maximum sensitivity (8–15 weeks after conception) corresponds with the major waves of renewal, proliferation and migration within the cerebral cortex of the brain. There is evidence of a linear, no threshold, dose–response relationship as judged by intelligence test scores of Japanese children who were in-utero at Hiroshima and Nagasaki. There appears to be very much less effect for exposures before 8 weeks and after 25 weeks of gestation. The risk in the sensitive period is estimated to be as high as 45 per 100 children per Gy of exposure.

The sensitivity of the embryo/foetus is known to be high for cancer induction as well. The best data come from the Oxford Survey of Childhood Cancers (OSCC) which was a case-control study of childhood deaths due to cancer in the period 1953–79 following irradiation in-utero. Early predictions (in 1956) that the study would reveal a doubling in the risk of childhood cancer (up to age 16) were not exactly confirmed by later studies. The present relative risk estimate is about 1.4–1.5 for the second two trimesters. There is, however, some controversy over the absolute risk, with both the OSCC and the Biological Effects of Ionizing Radiation Committee of the US National Academy of Sciences (BEIR III Committee) suggesting risk rates for cancer induction of 6 per 100 per Gy, and UNSCEAR in its 1977 report suggesting 2–2.5 deaths per 100 people per Gy. The main difference here derives from estimates of dose per X-ray film. The higher risk estimate is supported by the NRPB although it also maintains that only about half of these cancers will be fatal. These data are of some significance now, because of risks to the general public from environmental pollution and because they are applicable at low doses and dose rates. These data as sources of information for risk estimation and the setting of dose limits are explored in more detail in Chapters 2 and 3.

Effects of Radiation on the Immune System

The immune system

The immune response to an invading antigen (a foreign substance) is mediated through two cell types: the lymphocyte and the macrophage. It seems that the macrophage puts antigens into a form that is recognizable by lymphocytes. Lymphocytes have easily the best understood roles in the immune response. These are:

(a) recognition of the foreign substance (antigen);
(b) induction and regulation of the immune response;
(c) making the response very specific; and
(d) responsibility for immunologic memory.

Lymphocytes can be divided into two major groups on the basis of their function: thymus-derived (T) and bone marrow (B) cells. There is major cooperation between B and T cells, although to varying degrees.

The immune response can be divided into two broad categories: cellular and humoral. Cellular immunity, mediated by T cells, is the rejection of incompatible grafts, graft-versus-host disease and delayed hypersensitivity responses. The humoral response is that of antibody synthesis which is mainly the responsibility of B cells.

The T cells responsible for cellular immunity appear to have stem cells originating in the bone marrow that migrate to the thymus to mature. As they mature, the cell surface changes and they acquire the ability to distinguish foreign substances.

The B cells also undergo a maturation phase that may occur in the bone marrow. With the help of T cells, B cells differentiate (change) into plasma cells. These secrete antibodies and specific immunoglobulins that are able to attack foreign substances.

Once B and T cells are mature, they migrate to the peripheral lymph organs, i.e. the lymph nodes, spleen and Peyer's Patches (lymphoid sites in the gastro-intestinal tract). There is, however, a substantial circulation of lymphocytes, particularly T cells.

Small lymphocytes in the circulation are very radiosensitive. They are not a dividing population and thus have a rather anomalously high sensitivity to radiation-induced cell death. However, antigen-activated lymphocytes are much more resistant and plasma cells are incredibly radioresistant, being able to secrete antibody after doses in excess of 10 Gy.

Depression of the immune response after irradiation is mainly due to the radiosensitivity of small lymphocytes but can only be detected at quite large doses. Dose protraction or at reduced dose rate has much less effect on immune depression. Overall, evidence seems to indicate that large doses are required for functional effects to be manifest.

At high doses there is no doubt that the immune system can be effectively wiped out by radiation exposure, mainly because of extensive damage to lymphocytes. Doses that will completely ablate the bone marrow are used to prevent bone marrow graft rejection but normal levels of immunoglobulins are restored in a few weeks. At lower doses, the evidence for any functional effect on the immune system is very thin. Experiments with immunodeficient or immuno-suppressed mice have failed to reveal more sensitivity to irradiation in terms of increased cancer. This argues against suppression of the immune system being a significant factor in radiation carcinogenesis. Nevertheless, it is possible that in these mice some organ other than the bone marrow or thymus had taken over the role of immuno-surveillance. Even experiments on natural killer cells (those cells that are derived from large lymphocytes stimulated by certain substances released from cancer cells) have only shown a small effect of radiation on their production. However, most of these experiments have been carried out on adults, and more attention should perhaps be given to effects on foetal or neonatal animals. It does appear, for instance, that children who have received large in-utero exposures do have a temporary decrease in antibody competence.

At this point it is pertinent to examine the evidence for a slight augmentation or stimulation of the immune system by small doses of radiation. This, it is suggested, might occur because of radiation-induced injury of a sub-population of suppressor T cells involved in regulation of the response to an immune challenge. This overall phenomenon is known as hormesis.

Hormesis

The term hormesis was first used about 45 years ago to describe the stimulatory effects of extreme dilutions of toxic plant extracts on wood-decaying fungi in culture. It is in fact generally thought of as a botanical term and can be defined as 'the stimulus given to an organism by non-toxic concentrations of a toxic substance'. In the context of this book, it would cover the unexpected absence of radiobiological effect at very low doses. This is a subject that has been at the edges of experimental science for some time, and whereas there are plenty of theories, e.g. slight stimulation of the immune system, to explain the phenomenon should it be proved to exist, there are no sound empirical data. However, it *is* worth investigating the claims, although what exactly would be done by the dose regulatory authorities if good evidence of a hormetic effect were found is open to speculation. It would certainly *not* be a good reason to raise dose limits.

The hypothesis of hormesis for radiation has been with us for a long time, at least from the time of experiments on mouse longevity carried out in the late 1940s and early 1950s. Much of this earlier work suffered from rather variable disease control but, even in the early 1970s, the US National Council for Radiation Protection and Measurements in a publication on 'Basic Radiation Protection Criteria' could offer only a weak statement: 'In fact exposures at dose rates of about 0.1–0.5 rads per day have regularly resulted in an increase of the average life span in rodents, for reasons incompletely understood'. Even today, at lower doses, the same statement could be made, although the likelihood of any definitive animal experiments at very low doses being carried out is receding. Such experiments are difficult to perform with adequate controls, and are time-consuming and therefore expensive. Thus, we must look at the experiments that have been done and also at human experience, where it has been adequately recorded.

Experimental hormetic results

The first thing that obscures results on hormesis that have been published is scientific bias. There are two extreme views about the effects of low doses of radiation. One is that at low doses radiation is an essential biological nutrient for life. The other, the more conventional conservative view, is that any level of ionizing radiation may cause damage when applied to a biological system. Unfortunately, scientists in this field are hardly ever totally without bias in the *interpretation* of their results, even though their experiments have been performed accurately and with integrity. The most reliable data, therefore, probably come from experiments in which a hormetic effect was an incidental finding. Nevertheless, there are some rather compelling data, albeit at doses that are not that small, e.g. increased fertility in mice at doses of 100 μGy per hour, optimum growth of bacteria at 50–100 mGy per hour, increased growth of paramecia (single-cell organisms) at background doses, etc. Some of the data quoted most often concern effects that may not be beneficial, for example, increased body weight of mice exposed to chronic doses in the range 0.001–0.09 Gy per 8 hours. This weight increase occurred early in life and may be due to excess fat deposition, a doubtful benefit! There are now, no doubt, enough data on animals (and maybe even man) exposed to low levels of radiation to show that a slightly increased median life span is due to less early death accompanied by accelerated death rates later in life. Control groups do tend to have a higher variance for life span but always have the longest-living individuals.

The whole subject of hormesis was given an enormous fillip in 1985 when the Electric Power Research Institute (EPRI) in the USA organized a conference on radiation hormesis. Subsequently, twenty-two of the thirty-six papers presented at the conference were published in 1987 in the prestigious journal *Health Physics*. The ERPI conference report emphasized the concept that low doses of radiation induce repair enzyme production and also the fairly well-documented adaption of cell systems to continuous irradiation. There were also a number of human studies mentioned and these are probably the most fascinating and controversial. Much of this material is epidemiologically of doubtful statistical significance but some of the studies on inverse relationships with size of natural background are persuasive. The so-called 'healthy worker effect' also urgently needs explanation. This is the apparent (or real!) decrease in mortality from cancer of radiation workers as a group. This effect has been seen to be as high as 25 per cent when compared with national statistics in the UK and the USA. It is lessened by comparison with control groups matched for age, sex and socio-economic group, but has never been quite eliminated (this effect is explored in more detail in Chapter 2). Is this a beneficial effect of low doses of radiation? It is difficult to judge because mutagenesis and carcinogenesis are conceptual difficulties in the hormesis context. If low dose irradiation stimulates growth, then uncontrolled growth (cancer) should be stimulated as well. Maybe experiments with cells, etc. are inappropriate here and hormesis is, as has often been suggested, the response of the whole organism.

Perhaps one of the most interesting studies that may have a bearing on the low dose debate is a comparative epidemiological study being carried out in two areas of China that differ in background dose by a factor of 2.6 (5.5 mSv and 2.1 mSv per year for external plus internal dose rates). About a million person-years of experience in the study have been accumulated so far. Despite the differences in dose rates, the recorded cancer mortality rate is lower in the high dose rate area – corrected for age distribution, the mortality rates are 48.8 and 51.1 per hundred thousand person-years. However, the errors on these two figures are such that the opposite conclusion to a hormetic effect cannot be excluded. Apart from cancer mortality, a number of effects such as differences in growth rates, congenital malformations, hereditary disease, spontaneous abortion and neonatal mortality were not significantly different. However, stable chromosome aberrations were found to be related to dose rates. These studies are continuing.

In summary, the evidence for radiation hormesis is not that strong, and when put alongside the evidence, also fairly weak, for effects at low dose rates, is probably at best equivocal. However, there are plenty of convincing theories to explain such a phenomenon and I think it is as well to keep an open mind. If, or until, the phenomenon is demonstrated in a statistically significant manner, a conservative approach to dose limitation and control is the most favoured course. The attitude of those who regard this latter approach as restrictive on the nuclear industry and wasteful of valuable resources is a little odd to say the least.

2 Sources of Information on Radiation Risk

The Japanese Atom Bomb Surviors

At precisely 08.16 on 6 August 1945, the first atomic bomb was exploded in the bright morning sky above Hiroshima, Japan. This bomb, equivalent in blast power to about 15,000 tons of TNT, virtually flattened the city and immediately killed about 78,000 people, about 35–40 per cent of the population. Destruction was so great that one had to be more than 2.1 km from where the bomb landed to have a better than 50 per cent chance of survival. As if this hideous mayhem were not enough, three days later, just before 11.00, a slightly smaller bomb was dropped on Nagasaki. Partly because of the topography, fewer people (about 40–50,000) were killed, but the destruction was just as appalling as in Hiroshima.

These two bombs had the desired effect; on 10 August the Japanese surrendered and the Second World War effectively came to an end. However, for many thousands of Japanese this was just the beginning of the end. Not only did the blast and heat from these two bombs kill immediately, but the radiation released has wreaked a continuing death toll ever since. Gruesome as it is, the Japanese experience has provided the best information there is on the long-term effects of radiation exposure. This acquisition of data was made possible by a 1947 decision of the US National Academy of Sciences to set up the Atomic Bomb Casualty Commission (ABCC). In 1975 the ABCC was superseded by a joint Japanese–American organization, the Radiation Effects Research Foundation (RERF), which still follows the health records of survivors. The information that has come from studies of their health and, more significantly, their mortality has had an important bearing on our knowledge of radiation-induced cancer and has helped in the formulation of standards of dose limitation.

The major study of cancer mortality since the bombings in Japan consists of a follow-up of members of the Life Span Study (LSS) cohort. This consists of survivors from both cities who were alive in 1950. Infrequently since then, the total mortality and causes of death of individuals have been analysed and reported. The latest reports (published in 1988/89) cover the period up to 1985.

Apart from the inclusion of extra deaths since the last report (1978), the latest report contains several other important findings, some of which may trigger changes in radiation protection dose limits. An extra 11,000 survivors from Nagasaki have been included, bringing the LSS cohort up to more than 120,000. Of these, 29,000 have been classified as 'not-in-city' and are not used for dose–response analysis; 37,000 in a zero dose group are used as controls. The remaining 54,000 are divided into seven dose groups: 10–90, 100–490, 500–990, 1000–1990, 2000–2990, 3000–3990 and 4000+ mGy. These doses are assessed from dosimetry that was established in 1965 and slightly revised later. Based on this dosimetry (the so-called T65D system) the average dose was estimated to be 390 mGy in Hiroshima and 420 mGy in Nagasaki.

Clearly, from the point of view of attempts to establish reliable risk rates for cancer mortality, the dose that was received by each individual survivor is of some importance. But these doses are never going to be known precisely, although as a result of considerable effort the T65D system was developed by the USA to make *some* estimate. A simulation of the conditions of the Hiroshima bomb was made in the Nevada desert together with replicas of Japanese houses, and doses in them from nuclear bomb tests of similar yield were recorded. This presented problems because the Hiroshima bomb was a one-off gun barrel type and could not be remade. However, there was no immediate reason to suspect that the dosimetry was inaccurate until a marked difference was observed between the dose–response curves from the two cities for leukaemia induction.

In Hiroshima the leukaemia mortality curve was essentially linear with dose whereas in Nagasaki there was no increase at low doses, suggesting a threshold. The explanation offered for this difference rested on the fact that there were variations in the type and structure of the bombs; in particular, there was a much bigger neutron component at Hiroshima than at Nagasaki. Because of the high LET (and, therefore, high quality factor, see Chapter 1) of neutrons, it was suggested that neutrons posed the main hazard.

However, the dosimetry began to be seriously questioned in the early 1970s as a result of calculations done at Oak Ridge and Los Alamos laboratories in the USA. For several reasons it was realized that the neutron component at Hiroshima and, to a certain extent, Nagasaki had been wildly overestimated; maybe by a factor of ten. The reasons for this error were connected with the type of bomb; the structural material, which afforded more shielding than was expected; and the humidity in Japan, which attenuated the neutron flux by more than had been estimated. These revelations were followed by the setting up of a joint Japanese–American research programme to reassess the individual dosimetry at both cities and report on the biological implications. The new dosimetry system (DS86) was unveiled in 1986.

Apart from a reduction by a factor of ten in the neutron component at both cities, the gamma component was increased at Hiroshima and slightly reduced at Nagasaki. As well as reassessing the in-air dose, researchers considered it appropriate to look more closely at attenuation (shielding) factors in houses and by body organs. As complete an analysis as possible was carried out of each survivor's location and position at the time of the explosion. From this analysis organ doses were calculated.

Only 18,500 survivors yielded sufficient data to calculate their doses directly by the new dosimetry. For 57,500 people, the doses were calculated indirectly. The 76,000 people forming the two groups make up the DS86 sub-cohort of the LSS group for direct comparison with the T65D system. This leaves 17,000 of the original T65D group without a new dose assignment. This distinction is important because it was not immediately clear that the T65D and DS86 groups were different.

Because of the absence of the neutron component, the transmission or shielding factors differ by up to a factor of two. However, whereas the shielding factors for house structures increased, the shielding due to body organs and overlying tissues was reduced by an almost equal amount; i.e. they almost cancel each other out for some tissues. Another consequence of the new dosimetry is that the new dose–response data are not sensitive to the QF or Relative Biological Effectiveness (RBE – see Chapter 1) chosen for neutrons. This has always been a contentious point.

From about 1987 reports began to come from the RERF of changes in dose–response data in terms of cancer mortality. These included not only changes due to the DS86 dosimetry but also, and much more significantly, changes due to the inclusion of extra deaths since the previous reports in 1978. For both leukaemia and all cancers, there was more evidence under 4 Gy for a linear dose–response relationship, although it was still possible to fit a curve called a linear quadratic (see Chapter 3). The estimated risk of leukaemia under the new dosimetry increased by 48 per cent, but the overall increase in risk is only about 5 per cent. There were, however, much greater increases in risk for some individual cancers such as those of the lung (35 per cent) and breast (133 per cent). Nevertheless, these were offset by only small increases in risk for other cancers that are more common in Japan.

Thus, for solid cancers, changes in dosimetry in themselves had little influence on estimates of the cancer risk. Of greater significance was the increase in cancer deaths that took place as time passed. This necessitated a reappraisal of the risk projection models used to estimate lifetime cancer risk.

The increases can be seen in Figure 2.1 for all cancers; the significant decline in leukaemia cases after a peak in the early 1950s can also be seen, which is the reason why leukaemia is usually analysed separately. However, before looking more closely at these data and deriving risk rates for radiogenic cancer mortality from them, let us investigate the other sources of information from which risks have been assessed.

Apart from the survivors of the atomic bombings in Japan, there have been a number of groups of people who have been exposed to radiation in sufficient doses to show effects that could be used to estimate risk. The list of these groups is long and growing, with the Chernobyl disaster now adding to it. Patients treated with radiation for cancer and other diseases feature in the list. It is ironic that radiation used to treat cancer because of its cell-killing properties has also resulted in second cancers developing in other tissues.

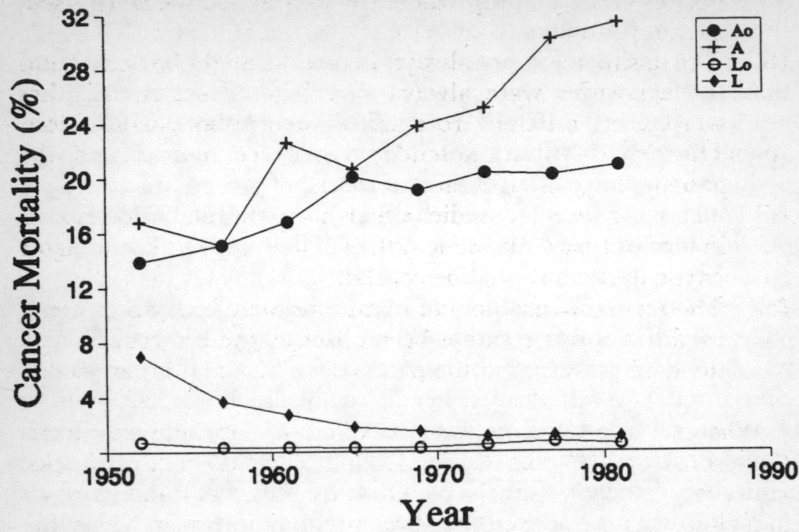

Notes: Curve A is for the zero dose group and Curve B is for the five dose groups down to 0.5 Gy. Curves A and B refer to all cancers except leukaemia and Curve A′ and B′ to leukaemia.

Figure 2.1 Cancer Deaths at Hiroshima and Nagasaki (1950–82).
Note: Expressed as a percentage of the deaths from all causes, plotted against the years in which they occurred. A and o are for the zero dose groups. A & L are for the five dose groups down to 0.5 Gy. A = All cancers except leukamia. L = Leukaemia.

Medical Exposures

Radiation exposure for medical treatment or diagnosis has been used extensively since the earliest days of this century, sometimes indiscriminately. There are thus a number of groups of people who have received doses (usually high) of radiation that have produced effects later. These groups are potentially of interest but their value in the low dose effect debate is limited by a number of factors:

(a) Generally they were ill with a medical condition that was thought to be susceptible to radiation treatment. This would not, of course, apply where diagnostic X-rays were used, even repeatedly.

(b) The dosimetry is not always as good as might be hoped and the exposures were always very heterogeneous; in other words, very different from uniform whole-body irradiation.

(c) The age distribution tends to be very limited and the patients have often been of one sex.

(d) There are very few studies which have good statistical power in terms of man-Sv and length of follow-up. The latter factor particularly may well be crucial.

(e) The recurring problem of inappropriately high doses being used has to some extent been offset by the heterogeneity of the dose pattern and the protraction (in time) often used.

When set alongside other sources of data on radiation effects, there is no doubt that medical data have their drawbacks. However, no data presently available are perfect and therefore we must look, at least for confirmation, at all information. Nevertheless it is significant that the latest report of UNSCEAR (1988) considers two medical studies – the ankylosing spondylitic series and the cervical carcinoma patients – and then effectively rejects them in favour of the data from the Japanese bomb survivors. Let us therefore briefly appraise the medical radiation epidemiological studies.

The spondylitics

Historically, the most important study has been the follow-up of the people treated for ankylosing spondylitis with radiation in the period 1935–54. Ankylosing spondylitis is a crippling and painful disease of the spine but it was found that X-irradiation of the spine with fairly large doses (2–12 Gy) was effective in reducing symptoms. This treatment was used at more than eighty centres in the UK with and without chemotherapeutic agents. The patients exposed to a single radiation treatment course have been followed with greatest care. However, data from even this group have never been good (Figure 2.2) with large confidence limits. The dose–response curves are erratic and consistent with a variety of different models including one in which risk did not vary with dose. Nevertheless, the incidence of leukaemia, which first drew attention to the possibility of late effects in these

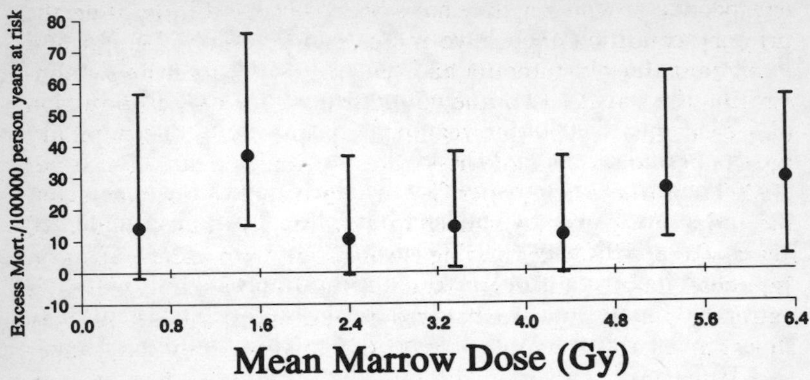

Figure 2.2 Excess Death Rates (With 90 per cent Confidence Intervals) from Leukaemia in Spondylitis Patients More Than Eighteen Months After First Treatment, According to Mean Bone Marrow Radiation Dose.

patients, was somewhat similar to the Japanese bomb survivors in that it followed a wave-like pattern of excess risk with time, peaking between three and eight years after exposure and declining thereafter. In the first twenty-five years after exposure, a fairly substantial increase of solid cancers in heavily irradiated organs was also seen but then these excesses declined. In the longer term (greater than 20 years) follow-up data are in no way similar to the Japanese in that the relative risk appears to have declined to about one for all cancers. These data seem now to be anomalous, especially since the risk of non-cancer deaths now exceeds the risk of cancer.

Children treated with radiation

There have been a number of radiation treatments for benign conditions in childhood. Those studied include:

(a) Treatment (epilation) for scalp ringworm (tinea capitis) – 10,842 children.
(b) Treatment for enlarged thymus – 2650 children.
(c) Treatment for tonsilar and nasopharyngeal conditions – 2578 children.

All these treatments have been to the head and neck, and irrespective of whether they have been successful in treating the primary condition they have increased the risk of benign and malignant thyroid tumours and leukaemia. Apart from demonstrating the sensitivity of the young thyroid for cancer induction (3.5 cases per 10,000 per year per Gy, mostly of the papillary type), these data also give risk rates for leukaemia (0.9–1.0 per 10,000 children per year per Gy) at relatively low dose rates and the induction of breast cancer in children irradiated under 10 years of age. This latter finding is consistent with observations on Japanese bomb survivors and, somewhat unexpectedly, is extremely important. The data from children exposed in this way show remarkable consistency with other studies for thyroid cancer and leukaemia. However, although the follow-up has been in excess of twenty-five years, which is adequate for leukaemia, it may not be long enough for solid tumours. The studies do emphasize the increased relative risk of children under 10 years of age for leukaemia induction.

The tinea capitis study has also shown an excess of brain cancers with doses in the range of about 1–1.4 Gy (to the brain). A risk estimate of about 3 cancers/10^4/year/Gy has been calculated.

Exposures in-utero

It was once fairly common practice to X-ray mothers during pregnancy when any potential problem was identified. The fact that this is practically never done now is partly because of the introduction of techniques such as ultrasound and partly because of the efforts of Alice Stewart.

In the 1950s, Stewart and her colleagues did a careful retrospective study of malignant disease in children, each child being matched with a healthy control of the same age and sex resident in the same area. Although there were some rather 'woolly' aspects to the original survey (for instance reliance on mothers' memories for evidence of X-ray exposure), it was claimed that antenatal irradiation doubled the risk of a child developing cancer before the age of 10. At the time (1956) this claim was quite controversial and several other studies were set up to test it.

The first large study was made in the USA by MacMahon and produced data that tended to confirm the extreme radiosensitivity of the unborn child. Subsequently (1970) Stewart reanalysed what was known as the Oxford Childhood Cancer Survey (OCCS) data and reported information on 7649 pairs of children born between 1943 and 1965. This analysis, covering a longer period, indicated a falling risk with time and a 40–50 per cent increase in childhood cancer associated with prenatal X-rays: i.e. the risk increased from about 2.2 in 100,000 to about 3.1 in 100,000. She found that of those who had died of cancer before the age of 10, 1141 had been X-rayed in-utero. Of an equal number of children who did not contract cancer, only 774 had been X-rayed. The actual dose the unborn child received has always been a contentious issue because X-ray methods were constantly evolving at that time and there was always some doubt about the actual number of films taken, and the dose per film. One of the better-estimated ranges is 2–4.5 mGy per film with 1–5 films taken.

Stewart's conclusions from these data attracted considerable media attention but were received coolly by the medical profession which seemed unable to comprehend that any risk could be associated with medical practices. However, several reanalyses of the data have suggested alternatives to radiation as the sole causative agent. One argument centres around the suggestion that problem pregnancies deemed to require X-rays produced children who were genetically, or otherwise, predisposed to cancer. These arguments have been refuted by consideration of the effects of X-rays on twins in-utero, where it had been the practice to use X-rays to confirm the presence of twins rather than for the treatment of any adverse condition of the mother. This line of reasoning was developed in 1974 by Dr Robin Mole, an eminent radiobiologist. Using the OCCS data, he noted that twins were five times more likely to be irradiated than singletons (as shown in Table 2.1), but that twins who were irradiated had a much higher cancer risk than those who were not.

The OCCS data are now generally accepted with few reservations. One concern has been the relatively large effect that Stewart *et al.* noted in the first trimester of pregnancy. Mole has pointed out that X-rays carried out in the second and third trimesters were generally for obstetric reasons and would have involved relatively small doses. By contrast, in the first trimester X-rays were for non-obstetric reasons (including fluoroscopies) and the doses less quantifiable but much larger. He concluded that not only are there uncertain estimates of foetal dose per X-ray film

Table 2.1 Effects of In-utero Irradiation in Singletons/Twins (Death Rates are per 100,000 Before 10th Birthday)

	Prenatal X-rays (%)	*Death rate in non X-rayed* leukaemia	other cancers	*Excess death rate in the X-rayed* leukaemia	other cancers
Singleton births	10	22	26	12	13
Dizygotic (unlike) twins	55	16	21	9	10
Monozygotic (identical twins)	55	0	11	29	18
All twins	55	12	20	14	11

made for obstetric examinations late in pregnancy but also that none are available for the conceptus or embryo in non-obstetric examinations in the first trimester. There is some support for this stance but it is difficult to resolve in a retrospective study.

One other inconsistency often cited is that the OCCS data do not fit in any way with the lack of in-utero effect in Japan in children who were irradiated, generally, in the third trimester of pregnancy. However, the atomic bomb survivors were subjected to high, cell sterilizing doses, but whether this resolves the argument is unclear.

From these data, the number of excess cases of cancer has been variously quoted in the range 4.5–6 per hundred per Gy. The NRPB has recently (1989) quoted excess rates of 2.5 per 100 children per Gy for leukaemia and 3.5 per 100 per Gy for solid tumours following exposures in-utero. Approximately half of these cases will be fatal. There is no doubt that the perceived greater sensitivity of the unborn child has led to fewer X-rays of pregnant women, particularly in the early stages. Prudence suggests that this is the correct action even if radiation is not the sole causative agent in the induction of cancer.

Cervical cancer patients

One of the largest studies of cancer produced as a result of medical radiation treatment has been the follow-up of women treated for carcinoma of the cervix. Treatment has involved radium implants and/or external beam therapy to the pelvis and lower abdomen. The radiation doses to the target area have been large (about 10–70 Gy) but the epidemiologists claim the dosimetry to peripherally related sites is good. However, the average follow-up has been fairly short so far. This study involves international cooperation between eight countries following a large sample of more than 182,000 women. The doses to organs near the tumour site are very large and, although fractionated, cell sterilization is clearly more important than cancer induction. In fact, the results are anomalous in several respects that may be concerned with dosimetry: for instance, the incidence of leukaemia is hardly detectable and, apart from lung cancers, no tumours are showing significant excesses. The stomach, which received an intermediate dose, showed no increase in cancer as was the case for breast cancer. The latter effect (or lack of one) has been explained by a radiation protective hormonal effect due to ablation of the ovary.

Together with the ankylosing spondylitics study, the cervical carcinoma study has the largest number of subjects and UNSCEAR in its 1988 report has considered these two and the A-bomb survivors as a special group for analysis. The comparison of derived risk rates lends no hope to the ideal of consistency (see Table 2.2).

Epidemiological Studies of Radiation Workers

The third group of people that could provide information on the long-term effects of radiation are those exposed as a result of their work. Studies of these people should be potentially the most apt and valuable because the doses received are at the level of most interest. However, the health follow-up of these groups is hampered by the very fact that the doses *are* small and that therefore the number of excess cancers induced is also small. In addition, the length of follow-up, which is of crucial importance, is not yet really long enough. Nevertheless the data that have been accrued bear some closer scrutiny because controversy has arisen over extravagant claims made using these studies as evidence.

The Hanford study

No epidemiological study of radiation effects has created more

Table 2.2 Cancer Risks in Three Populations (Excess Cancer Deaths per 10,000 People per Year per Gy)

Cancer	*A-bomb survivors*	*Spondylitics*	*Cervical cancer patients*
Leukaemia	2.94 (2.43–3.49)	2.02	0.61
All cancer except leukaemia	10.13 (7.96–12.44)	4.67	*

Note *not calculable

controversy than that carried out by Thomas Mancuso, Alice Stewart and George Kneale on the employees at the Hanford works in the USA. The Hanford works, situated near Richland in Washington State, cover an enormous area of nearly 1950 sq km. The plant, which has been a reprocessing and plutonium extraction facility and is also the site of a large radioactive waste dump, went into full production in 1944. About 44,100 people, mostly white males, have worked there and more than 35,000 have been monitored, mostly with film badges for external radiation, but also for intake of radioactive materials.

In 1964, an epidemiological study of workers' deaths was begun for the US Atomic Energy Commission by Mancuso. In 1977, before he could publish his results, his funding was terminated and the contract given to three other institutions largely supported by the DoE (Department of Energy). This action alone caused some eyebrows to be raised, particularly when the results of the Mancuso study were finally published later in 1977. By this time, Mancuso had the collaboration of a British epidemiologist, Alice Stewart, and a statistician, George Kneale. These three then published the first of a series of lengthy papers on the health effects of work at Hanford in relation to radiation exposure. The earlier papers, particularly the first, made controversial claims. The interest of the radiobiological community worldwide was therefore alerted and more than fifty papers have now appeared on the epidemiological data.

In their 1977 paper, Mancuso, Stewart and Kneale published results that showed a strongly positive association of cancer with radiation. They concluded that the risk of death from radiation-induced cancer was about ten times greater than 'conventional' estimates. These data were derived from deaths of Hanford workers (4000) up to 1973 (as identified by the nationwide system of social security numbers, through death benefit claims) correlated with their film badge dose-records. They reported that the cumulative mean radiation dose (CMD) received by those who died of cancer (2.10 rem/0.021 Sv) was higher than that received by those who died from other causes (1.62 rem/0.016 Sv), but they noted that surviving workers, anomalously, often had the largest radiation doses. They extended these analyses to a number of specific cancers and deduced doubling doses, i.e. doses that would double the spontaneous incidence of the cancer. These were the most controversial of all their results, e.g. 0.8 rad (0.08 Gy) for bone marrow cancers (leukaemia), 7.4 rad (0.074 Gy) for pancreatic cancer, 6.1 rad (0.061 Gy) for lung cancer and 12.2 rad (0.12 Gy) for all cancers. Significantly, leukaemia had occurred much less than was expected.

The statistical methodology and conclusions of this 1977 paper were heavily criticized. There followed a rebuttal of these criticisms in 1978 and then a third, more sophisticated analysis in 1981. In a sense the study was weakened by the last two papers but many of the criticisms had been heeded and quite dramatic changes were made. For instance, the doubling dose for all male cancers rose by nearly a factor of three to 33.7 rad (0.34 Gy). However, the cumulative mean dose (CMD) concept was retained (rather than the more conventional standard mortality ratio, SMR) but the analysis was now controlled for several possible biases, e.g. sex, age at death, date of death, exposure period, gross level of internal radiation (not quantitatively determined). Doubling doses from the 1978 paper are shown in Table 2.3. It should be noted that the confidence intervals are so wide that it is doubtful whether any definite statistical conclusion could be made from these results.

The third paper was written as a direct answer to the criticisms voiced over the first two but, by general consensus, many of the main objections were avoided although the CMD approach was defended on the grounds that the actual number of deaths is not known (some could not be traced). This remains a considerable deficiency.

Stewart *et al.* were attacked for not using the current US mortality statistics (they used 1960 data). This is important because trends of death from specific causes change with time. For

Table 2.3 Doubling Doses for Hanford Workers (1978 Study)

Cancer type	*Doubling dose (rad)*	*95% confidence limits*
Male		
Myeloma and myeloid leukaemia	3.6	2–10
Lung cancer	13.7	7–29
Pancreas, stomach, large intestine	15.6	7–55
All cancers	33.7	15–80
Female		
All cancers	9	3–∞

instance lung cancer mortality rates increased by over a factor of three over the period of the Hanford study, 1944–77, and they accounted for 29 per cent of male cancer deaths in the study. Also no allowance was made or control used for smoking, which must have been a significant contributor to the lung cancer mortality (it was very closely associated with emphysema which suggests a reflection of smoking habits).

Many of the subsequent reanalyses of the data presented by Mancuso, Stewart and Kneale agreed that only excesses of cancer of the pancreas and multiple myeloma were statistically significant, but all commented on the odd distribution of these types of cancer and the dose distributions. For instance, the distribution of doses actually received are so 'skewed' that the use of a mean seems highly doubtful: of the 11 men who died from multiple myeloma, 8 received less than 1 rad (0.01 Gy) and the other 3 account for 98 per cent of the dose, their doses being 35, 29 and 20 rad (0.35, 0.29 and 0.2 Gy) respectively. Thus, to attribute 9.7 deaths to radiation-induced bone marrow cancers seems unreasonable. Similarly, 2 of the 31 deaths due to pancreatic cancer account for over half the dose and 1 case of female breast cancer accounts for 14.4 rad (0.14 Gy) out of a total of 16.8 rad (0.17 Gy) for all cases. The means are clearly not good representatives of the true situation.

In addition, the excessive influence of the isolated large doses undermined the credibility of the doubling doses that were calculated. A doubling dose less than the background dose (0.8 rad for bone marrow cancers in the first report, subsequently revised to 3.6 rad) would imply a negative value for natural incidence! In addition, some age-related doubling doses were given that defy belief. For example, the doubling dose for reticulo-endothelial system (RES) neoplasm was given as 30 rad (0.3 Gy) at 45 years of age but only 0.1 rad (0.001 Gy) at 70 years. The doubling dose before 45 was apparently infinite but, in general, for other cancers, the studies deduced low sensitivity between 25 and 45 years of age with no statistical justification.

Lastly, one recurring criticism has been the use of a proportional mortality analysis. This, it is argued, has the disadvantage that a decrease in one cause of death produces an apparent increase in another. The analysis is therefore extraordinarily sensitive to changes in the distribution of deaths by cause. This was one of the reasons why Alice Stewart's testimony to the Windscale Inquiry (see Chapter 4) in 1977 was found to be so unconvincing by the Inspector (Mr Justice Parker), his actual words being: 'Although I found Dr Stewart's interpretation of the Hanford data unconvincing, I should perhaps stress that I have no doubt about either the importance of such data or the desirability of accumulating data of the same nature about radiation workers in the United Kingdom'.

Since that time, fourth (1981) and fifth (1984) papers have come from the Mancuso, Stewart and Kneale team, this time using more conventional statistical methods. The results still show the anomaly that radiation exposure apparently increases longevity, but their findings are now in broad agreement with their previous combatants. There have been some mild disagreements over whether local doctors in the Richland area would be biased towards overdiagnosing, but generally only multiple myeloma and possible pancreatic cancers show any statistically significant excesses in the population. It is noteworthy that this agrees broadly with a very large study of this population done by the General Accounting Office in the US (reported in the Comptroller General's Report to Congress for 1981).

Alice Stewart's team now seem to have moved on to other problems but the Hanford radiation worker population is still important and should, in time, provide some useful information about experience at low chronic doses. The latest paper (1989) by Ethel Gilbert *et al.* from Pacific Northwest Laboratory, Richland has updated the data to 1981 and still has an excess of multiple myeloma as an outstanding finding.

Overall, the 'Hanford experience' has demonstrated that the use of new methodologies for a study with weak associations is ill advised from both a scientific and political viewpoint. The constant backing down that the Mancuso, Stewart and Kneale team were forced to make certainly reduced the value of a potentially very important study. However, this study has now been followed by several others on large populations of workers that will, either singly or taken together with the Hanford study, provide a vast database with which to evaluate the risks of low-level exposure. Prominent among these are three studies from the UK that could have been prompted by Parker's words (recorded above) after the Windscale Inquiry.

UK studies

The first of the UK studies (1985) looked at the mortality of employees of the United Kingdom Atomic Energy Authority (UKAEA) between 1946 and 1979. This was followed in 1986 by a study of workers at BNFL's Sellafield site and then in 1988 by a study of employees of the Atomic Weapons Research Establishment (AWRE) at Aldermaston. All three studies used conventional statistical methods based on SMRs (standardized mortality ratios) and looked for associations with absorbed radiation dose. It is noteworthy that the 'dose' recorded for this purpose was the external (penetrating) dose recorded by film badges. Although the possibility of intake of radioactive materials was 'flagged', no internal doses were calculated for assessment of total risk. This must be taken as a considerable deficiency in these studies as the risk of radioactive intake is recognized as high at Sellafield and AWRE at least.

The overall message from these three studies was that there was little conclusive evidence of an association between radiation exposure and cancer mortality. However, none of the studies had the statistical power, in terms of man-Sv, to exclude a range from a protective effect of radiation up to a risk that was evaluated as up to more than fifteen times the ICRP's cancer risk factors. Inevitably the top end of this range was seized upon by the media, and some irresponsible alarmist articles were written.

Minor variations in the studies were highlighted in the various publications. For example, in the UKAEA and AWRE studies, prostatic cancer clearly showed some increases with increasing exposure. In both the Sellafield and AWRE studies, the effect of a lag or latent period (i.e. removing the dose from the record in the

10–15 years before death from cancer on the grounds that this did not contribute to the cancer's initiation) was tested and did increase the likelihood of association with radiation exposure. In the Sellafield study, there were 'not significant' excesses of deaths from myeloma (7 observed, 4.2 expected) and prostatic cancer (19 observed, 15.8 expected).

These studies clearly demonstrated the need for complete and accurate data on mortality, and accurate diagnosis on death certificates or even possibly of registration of cancer. Also, of course, the average follow-up (shown in Table 2.4) was nowhere near long enough to ensure statistical power. This follow-up may need to be in excess of fifty years; for instance, two-thirds of the survivors at Hiroshima and Nagasaki are still alive but showing increasing excess rates of cancer more than forty-four years after their exposure.

The mortality in some of the more comprehensive studies carried out in the UK and the US is shown in Table 2.5 as a standardised mortality ratio (SMR); the number of deaths observed in the workforce expressed as a percentage of that expected on the basis of national mortality rates. Apart from the fact that in some cases the SMR is based on very few deaths, the

Table 2.4 UK and US Epidemiological Studies

Study population		*Deaths*	*Period of study*	*Average length of follow-up*	*Person-dose man-Sv*
UKAEA	39,546 (20,382 radiation records)	3,373	1946–79	16 years	660
BNFL	14,327 (10,157 radiation monitored)	2,277	1947–75	21.6 years	1,259
AWRE	22,552 (9,389 radiation workers)	3,115	1951–82	18.6 years	73.6
Hanford	44,100 (>35,000 monitored)	7,249	1944–81	16–20 years	831

Table 2.5 Mortality Rates from US and UK Studies

	SMR (number of deaths)		
	All causes	*All cancers*	*Leukaemia*
USA			
Hanford	79 (7249)	85 (1603)	71 (52)
Portsmouth Naval Dockyard	89 (4762)	94 (977)	94 (39)
Rocky Flats Weapons Plant	64 (334)	75 (79)	117 (5)
United Nuclear Corporation	82 (216)	85 (51)	84 (2)
Oak Ridge National Lab	73 (966)	78 (194)	149 (16)
Pantex Weapons Plant	72 (269)	60 (44)	128 (4)
UK			
UKAEA	74 (2856)	79 (827)	123 (35)
Sellafield Plant	98 (2277)	85 (572)	63 (11)
AWRE	77 (3115)	82 (865)	74 (20)

most important observation from Table 2.5 is that the SMRs are mostly substantially below 100. This phenomenon is often termed the 'healthy worker effect', and although many explanations have been given of this, none are totally convincing.

The healthy worker effect

The healthy worker effect is an important feature of the Hanford data and most other epidemiological studies of radiation workers. As can be seen from Table 2.5, it would appear that radiation workers are healthier (i.e. die less frequently, up to 20 per cent less, from the stated causes) than the general population. Various reasons have been postulated for this. For instance, it has been observed that the health of people who work steadily for one employer can be influenced by pre-employment medical screening, training, job preference, health awareness, routine medicals and the socio-economic benefits of steady employment. As a result, they tend to have better health and greater life expectancy than the general public. This is easy to understand because the general public would include those who are chronically ill. It is, however, difficult to know how this would effect the overall incidence of cancer unless long-term workers live longer, die less of non-cancer causes and that therefore a greater proportion of their deaths could be from cancer. This could lead to a false connection with cancer, although generally it is not seen to do so.

The healthy worker effect remains an enigma because it is difficult to eliminate. At Sellafield the selection of a control group, standardized as far as possible for age, sex and socio-economic group, was shown to be of vital importance. Thus the Cumbrian population was a better group than national statistics. However, this phenomenon may well be obscuring some effect or else it has been suggested that it could even be an example of hormesis (see Chapter 1).

Comparison of worker epidemiology with ICRP predictions

One of the main objectives of studying the health effects of radiation in workers is to compare the magnitude of the risk of cancer with populations exposed at high doses. The ICRP has estimated the risk of all-cancer mortality to be about 100 cases per million people (100 in 10^6) per 10 mSv exposure. If those risks are assumed to be spread over twenty-five years after a ten-year latent or lag period, these estimates correspond to annual excess risks of 4 per 10^6 people per 10 mSv for all cancers and 0.8 for leukaemias.

In the first two UK studies, the results are expressed in such a way as to enable a direct comparison to be made with ICRP. The annual risk for all cancers in these two studies is 17 cancers per 10^6 people per 10 mSv. The confidence intervals (i.e. the range known with 95 per cent confidence) are very large: −22 to 60 in the AEA study and −30 to 70 in the Sellafield study. On the face of it, the risks could be assessed as about four times that quoted by ICRP, but with such large confidence intervals it would be unwise to make any definite conclusion. In addition, if the US studies are included the risks are slightly lower but the uncertainty intervals even greater. However, such assessments are being made by independent epidemiologists and as the follow-up increases the statistical power goes up. The industry cannot afford to be complacent even though the health effects seen now are a legacy of bad practices in the past.

Summary

This chapter has briefly explored the sources of information that have been and are being used to estimate the risks of low-level radiation exposure. It has concentrated on the more important (in terms of statistical power) studies and therefore has not touched on more controversial studies in which often extravagant claims have been made from relatively thin data. No apologies are offered for this.

The studies discussed fall broadly into two groups:

(a) Populations exposed at high doses and high dose rates. These include the survivors of the bombings in Japan and also several groups of patients treated with radiotherapy.
(b) People exposed to low doses over a long time. These are mostly radiation workers in the nuclear industry but there are data from other groups of workers, e.g. US and UK radiologists that (potentially at least) could contribute significantly to the debate.

The information collected from all these groups has provided evidence of long-term health effects, in particular incidence of cancer. Thus, to be useful, the epidemiology has had to be carefully controlled, free of bias and the follow-up lengthy. It can be concluded from the data presented that there are really few, if any, studies that fit these criteria. In these circumstances all that can be said is that each of them provides *some* small pieces of the jigsaw.

Potentially the worker studies should provide the best clue to long-term health effects because complications such as dose and dose rate extrapolation models (see Chapter 3) should be unnecessary. Such studies should clearly be pursued with the utmost vigour. However, at present none of them have the length of follow-up required. In the same way the medical studies are of limited value because of lack of follow-up, selection bias and inhomogeneity of dose.

This leaves the study on the Japanese bomb survivors as the most robust for risk rate estimation. This has been recognized by UNSCEAR in its 1988 report, although the uncertainties implicit in the models used are also recognized (see Chapter 3). As will be seen from Chapter 3, the cancer risk rates that may be derived from this population are now uncomfortably high and, if confirmed by longer follow-up in worker studies, would represent an intolerable situation.

Note: The data discussed in this chapter has been exclusively from low LET radiation. The epidemiological data from groups exposed to high LET radiation, e.g. radon and its progeny in mines, are included in Chapter 6.

3 Risk Estimation and Dose Limitation

Historical Introduction

When Wilhelm Roentgen discovered X-rays in 1895 there was no particular reason why he should regard them as dangerous. When immediate physiological changes did not seem to occur after exposure to the new rays, some thought them safe or even beneficial, like sunlight. After all, light from the sun was known to pose no special threat unless exposure was excessive. The reddening of the skin that resulted a few hours later from such exposure was entirely predictable and could be controlled. However, some X-ray workers began to suspect that ulceration and scarring of the hands, although occurring some time after exposure, was caused by X-rays.

The earliest reports (1896) of biological damage were of eye irritation which, ironically, may have been due to the ultraviolet rays generated from fluorescence and not directly from exposure to the primary X-rays. However, reports of skin damage were becoming more frequent and in 1896 the eminent physicist Joseph Thomson carried out an experiment on his own hand. He deliberately exposed the little finger of his left hand to an X-ray tube intermittently for several days and carefully recorded the painful effects that followed. By 1900 most scientists in the field were aware of the hazards, and indeed the doses they were receiving were enormous by today's standards. In 1902 one of Thomas Edison's assistants, who tested X-ray tubes, contracted cancer and died. His death was attributed by contemporaries to his work.

Although many scientists were eager to exploit the new tool, there were those who sounded a note of caution. Even at this early stage lead-shielding collimators and higher-voltage X-ray tubes were suggested as a way of reducing the dose to a patient. The first formal statement on radiation protection came from the German Roentgen Ray Society in 1913; it recommended lead shielding for X-ray equipment to protect the operator from leakages. Two years later the British Roentgen Society suggested the establishment of safety regulations, but the First World War intervened and nothing was done. Radiological safety at that time was deemed to be a comparatively minor problem.

After the First World War interest in radiological protection was revived when it was perceived that radiographers and radiologists were dying, some of leukaemia, as a result of their occupational exposure. The ninth of these, a prominent radiologist at Charing Cross Hospital in London, died at the age of 36 in 1921. His death belatedly precipitated some action, and the British X-ray and Radium Protection Committee was set up. Its recommendations were presented to the First International Congress of Radiology in London in 1925 and subsequently adopted by the Second Congress in Stockholm in 1928. The French Academy of Medicine was at the same time making similar recommendations on safe working practices. As a result of international discussion at Stockholm, the International X-ray and Radium Protection Committee (IXRPC) was formed.

Thus, judged by the effects seen in those who may now be considered martyrs to medical radiology, an appreciation of the hazards of radiation had developed. This awareness was translated, qualitatively at least, into recommendations such as:

> The X-ray operator should on no account expose himself to a direct beam of X-rays . . . An operator should place himself as remote as practicable from the X-ray tube . . . The X-ray tube should be surrounded as completely as possible with protective material of adequate lead equivalent.

Dose control was achieved by limiting working time, e.g.

(a) not more than seven working hours per day.
(b) not more than five working days a week. The off-days to be spent as much as possible out of doors.
(c) not less than one month's holiday a year.

The recommendations were not really quantitative at this time, although with the setting up of the International Commission on

Radiation Units and Measurements (ICRU) at Stockholm in 1928, the Roentgen was adopted as the official unit of X-ray measurement. The newly formed International X-ray and Radium Protection Committee (IXRPC), anxious to establish some way of quantitatively limiting exposure, first agreed on a 'tolerance level'. This was adjudged to be about that amount of radiation that was one-hundredth that required to produce erythema (reddening of the skin). This exposure was not to be exceeded in one month. However, the 'erythema dose' was not (nor could it be) defined other than within the range 100–700 Roentgen, and so this dose limitation was incredibly vague.

Three points arise from these early attempts at dose limitation:

(a) a threshold, below which effects were not manifest, was assumed;
(b) the radiologist/radiographer was singled out for protection. Although collimation of X-ray beams had been introduced, real consideration of patient protection followed later (and even today could still be improved);
(c) the doses that were considered 'tolerable' were large by current standards, e.g. the exposure rate considered tolerable in 1934 was 0.2 Roentgen/day, which is roughly equivalent to ten times today's worker dose limits.

Very little effort had been put into medical surveillance of those irradiated, but in 1931 the IXRPC recommended regular medical and blood examinations to determine 'the acceptance, refusal, limitation or termination of employment'. In 1935, the US Advisory Committee on X-rays and Radium Protection recommended a lower limit of 0.1 Roentgen/day, but no changes in their exposure limits were recommended at the final meeting of the IXRPC before the Second World War in 1937.

The IXRPC next met in 1950 at the 6th International Congress of Radiology and changed its name to the now familiar International Commission on Radiological Protection (ICRP). There had been a great increase during the Second World War in not only the number of sources of irradiation, but also in the number of people working with and exposed to radiation, and the need for control was obvious. This was particularly relevant in the production and use of radioactive materials. This expanding industry and research had followed the initial discovery of activation by neutrons in 1934 and the later production of fission products from the fissioning of uranium. It had become clear in the 1930s and 1940s that the use of radium (in particular) and X-rays was burgeoning and not just for medical therapy and diagnosis. For example, fluoroscopy in shoe-fitting (the 'pedascope') was popular and X-rays were also used to remove facial hair in beauty clinics. Radium-containing waters were common treatments for a variety of conditions. X-rays were being used in medicine with little regard for long-term consequences; for instance, children with fungal infections of the scalp such as ringworm (tinea capitis) were epilated with X-rays as part of their treatment and a number of benign skin conditions were treated with superficial irradiation. The use of radium in the luminous paint applied to clocks and watches was also causing problems. In the early days of the use of this material, it had been the practice to apply the paint by hand using a small paintbrush. This was tipped in the lips and appreciable quantities of the radium were ingested by the operatives (mostly women).

A number of these patients, workers or members of the public are now suffering, or have died from, cancers that can be attributed to careless use of X-rays or radioactive materials. However, it is quite surprising how long it was before these practices were banned. Even in 1950, the newly formed ICRP made no statement about them but did reduce the recommended maximum permitted dose for radiation workers by a factor of 3 to 0.3 Roentgen/week. The reason given was that the previous limit of 1 Roentgen/week was 'very close to the probable threshold for adverse effects . . .' This emphasis on a threshold was odd because at the time there was beginning to be considerable awareness of the possible genetic effects of radiation and a threshold for such damage seemed unlikely. Nevertheless the concept of a threshold for some effects – those termed non-stochastic effects – that are known to occur at higher doses is reasonable and has subsequently been demonstrated.

In the 1950s the late effects of the atomic bombings in Japan had just begun to reveal what had already been seen in others decades before. In addition, programmes of nuclear weapons testing had been embarked upon by the USA, the Soviet Union, and the UK. The long-term effects in Japan have become the single most important source of data on the long-term risks of radiation exposure. With the passage of time and therefore length of follow-up, and the reassessment and refinement of dosimetry and careful analysis of health records, the data from Japan are now the main basis for the establishment of dose limitation standards. The other major epidemiological studies, particularly groups exposed for medical purposes, are now more often used as confirmation (or not!) because of bias and lack of statistical power. However, what progress was made in dose control in the 1950s and 1960s?

The International Commission on Radiological Protection

The newly formed International Commission on Radiological Protection (ICRP) was initially quite active. It set itself up as a Main Commission of a chairman and twelve other members of whom not less than three and not more than five could be changed before each four-yearly International Congress of Radiology. Four standing committees were set up to review all published evidence pertinent to radiation protection and to advise the Main Commission. Currently the Commission and these four committees involve seventy-five scientists from twelve countries. The ICRP, when first set up, was quick to emphasize its international character and that its role was solely to make recommendations. Nevertheless, most governments look to this body as the first and final arbiter in formulating their national radiation protection regulations.

The ICRP membership is seen by pressure groups as a 'self elected elitist society' and it has indeed tended to be drawn from governmental nuclear institutions and medical radiology. However, although it is difficult to see from where else experts with a sufficient breadth and depth of knowledge could be found, there is no doubt that the appointment of scientific representatives from labour unions would have helped its image without damaging its integrity. Certainly the ICRP was slow to act but in the 1950s several steps forward were made (see Table 3.1)

The concept of a 'tolerance dose' lived on for a few years, but by 1955 the threshold was thought to be too close to the dose at which effects occurred. In addition, it was thought that the genetic effect without a threshold was of overriding importance and therefore the dose limit was reduced successively in 1955 and 1959. ICRP 2 was also published in 1959, consisting of the report of Commit-

Table 3.1 UK Occupational Whole-body Dose Limits

1951	0.5 R/week
1955–9	0.3 R/week (200 R in a lifetime, averaging 5 R/year)
1959–77	5 (N–18) rem/year (3 rem/13 weeks)
1977–present	50 mSv (5 rem)/year

Note R is the old unit of exposure, the Roentgen; N = age

tee 2 on doses from radioisotopes taken into the body. This large tome provided data on the maximum permissible body burdens and maximum concentrations in air and water of all known radioisotopes. The basis of these recommendations was the 'radium standard', which had been recommended by the US National Bureau of Standards Advisory Committee in 1941. This suggested a maximum residual body burden of 0.1μg of radium. The annual dose this provided to the bone, 15 rem, was used as a limit for organs other than the bone marrow and the gonads for all other radioisotopes. There was *no* attempt to justify this limit in terms of risk as no risk estimates were available and the concept of a threshold or tolerance dose had only just been seriously questioned. It is therefore worth quoting verbatim ICRP's justification of its dose limits introduced in 1959.

> Estimates made by the different national and international scientific bodies indicate a per capita gonad dose of 6–10 rems accumulated from conception to age 30 from all man-made sources would impose a considerable burden on society due to genetic damage, but that this additional burden may be regarded as tolerable and justifiable in view of the benefits that may be expected to accrue from the expansion of the practical applications of 'atomic energy'. There is at present considerable uncertainty as to the magnitude of the burden and, therefore, it is highly desirable to keep the exposure of large populations at as low a level as practicable, with due regard to the necessity of providing additional sources of energy to meet the demands of modern society. A genetic dose of 10 rems from all man-made sources is regarded by most geneticists as the absolute maximum and all would prefer a lower dose. In some countries the genetic dose from medical procedures has been estimated to be about 4.5 rems. Therefore, if the limit for the genetic dose from all man-made sources was set at 6 rems, the contribution

> from all sources other than medical procedures, would be limited to 1.5 rems in these countries. This would impose unacceptable restrictions on these countries. Accordingly as a matter of practical necessity the Commission recommends that medical exposure be considered separately and that it be kept as low as is consistent with the necessary requirements of modern medical practice . . . In view of these considerations the Commission recommends a limit of 5 rems for the genetic dose from all man-made sources of radiation and activities, except medical procedures.

With this statement, the ICRP was already suggesting, even in the absence of quantitative evidence of long-term hazards, that there should be some trade-off between the risks and benefits of nuclear power. However, it seemed to be rather more concerned with what could be achieved rather than relative safety, and by excluding medical exposures this principle was assured. If medical exposure had not been excluded the expansion of the nuclear industry would have been severely held back, especially as medical per capita doses have not changed substantially in the last thirty years.

The ICRP also recommended, or suggested, a formula for relating age to total exposure (5(N-18)) that theoretically would allow 12 rem/year if the dose rate maximum of 3 rem/quarter was enforced. Nevertheless, one must hope or assume that repeated exposure of 3 rem/quarter would have been the subject of *some* investigation. These ICRP recommendations were reviewed several times in the decade without being changed materially; the continued support for the 5 rem annual limit being because 'the Commission believes that this level provides reasonable latitude for the expansion of atomic energy programmes in the foreseeable future' (ICRP 1966). However, in 1977, the ICRP felt able to justify its dose limits in terms of quantified risk. Its Publication 26 of that year is a watershed in the history of radiation protection.

The 1977 recommendations of ICRP

The recommendations issued by the ICRP in 1977 in Publication 26 (shown in outline in Table 3.2) were important for a number of reasons although they did not advocate any major revision of the dose limits. What they did do was to attempt a justification of the dose limits in terms of risk and to introduce general rules of radiation protection for workers and members of the public.

In reply to subsequent criticisms of the level of protection afforded by the numerical limits recommended in 1977, the ICRP has often fallen back on its 'basic tenets of radiation protection' to argue that radiation doses should be kept as low as possible. These basic tenets were:

(a) No practice should be adopted unless its introduction produces a positive net benefit;
(b) all exposures shall be kept as low as reasonably achievable, economic and social factors being taken into account (the ALARA principle);

Table 3.2 The 1977 ICRP Scheme of Dose Limitation

Dose Equivalent Limits

(a) *Non-stochastic effects* – where the severity of the effect varies with the radiation dose, and where they may be a threshold (e.g. erythema, cataract). The *aim* is to *prevent* these effects which, generally, are manifest at relatively high doses.

The *limit* to *individual tissue* is:
0.3 Sv (30 rem) per year for the lens of the eye
0.5 Sv (50 rem) per year for all other tissues.

(b) *Stochastic effects* – where the severity of the effect does not depend on the dose delivered, but the *probability* of the effects occurring does (e.g. cancer, genetic effects).

The *aim* is to *limit* the probability of these effects to levels deemed acceptable.

The *limit* for *uniform* irradiation of the whole body is

for designated radiation workers	50 mSv (5 rem) per year
	= 1 mSv (100 mrem) per week
	= 25 μSv (2.5 rem) per hour
for *other workers*	15 mSv (1.5 rem) per year
for *public*	5 mSv (0.5 rem) per year

The limit for *non-uniform* irradiation of the body

$$\Sigma_T W_T H_T \leqq 50 \text{ mSv (5 rem) in a year}$$

(c) the dose equivalents to an individual shall not exceed the limits recommended for the appropriate circumstances by the Commission.

The Commission stated that its recommendations were designed to prevent the occurrence of non-stochastic effects and to reduce stochastic effects, e.g. cancer, to an acceptable level (see Figure 3.1). Clearly there were some emotive and ill-defined words in these recommendations that have led to confusion and criticism. For instance, the terms 'reasonably achievable' and 'acceptable risk' are open to a number of interpretations.

Nevertheless the ICRP felt that it had been able for the first time to back its recommendations for dose limits for radiation workers with data on implied risks and felt these needed justification. Its basic philosophy was to equate the risks of death at the dose limits with those experienced in other industries. It

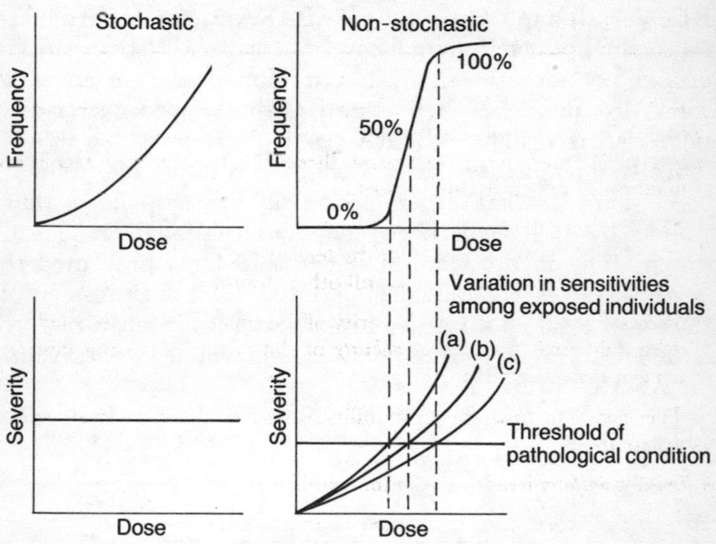

Figure 3.1 Stochastic and Non-stochastic Effects of Radiation Doses.

Note: The risk of stochastic effects (e.g. cancer) increases in probability with dose with no threshold. The severity of a stochastic effect is not related to dose whereas it is (with individual variation) for non-stochastic effects. Non-stochastic effects only occur above a threshold dose.

decided that an annual risk of death for a worker of 1 in 10,000 was acceptable. The risk in all manufacturing industries at that time was about one-third of this and it is a little better now (see Table 3.3). Exposure at the recommended dose limit (50 mSv/year) was five times this: 1 in 2000/year, which is also about five times the general risk of death in a traffic accident (see later for derivation of risk rates). Clearly, exposure at the dose limit would be seen by everybody as being intolerable. Therefore, this was explained in ICRP 26 (which equated the risk at the dose limit to 'risky' industries) and a hope was expressed that doses would be distributed such that most workers received a dose below 5 mSv/year (at which dose the risk could indeed be considered 'acceptable', i.e. 1 in 20,000/year). It was to be left to employers to ensure that these limits would be applied by diligently following the ALARA or optimization principle. However, no tests of the efficiency of ALARA were suggested and none has been used in the UK, so that in practice the principle is rather nebulous. The ICRP later attempted to explore ALARA in cost–benefit terms but the values put on human life were so disparate – between a few thousand pounds to well over a million pounds even in one country, let alone between countries – that the exercise was heavily criticized. The ALARA principle is so vague as to be practically unenforcable and is a good example of the lack of appreciation of radiation protection on the shop floor, that is exhibited by ICRP. Perhaps some international trade union representation on the Commission would have prevented this. Nevertheless, it is worth recording that at least one prosecution – of BNFL in 1983 as a result of releases of radioactive materials into the Irish Sea – has been pursued, and won, for contravention of the ALARA principle.

Table 3.3 Annual Industrial Fatal Accident Rates in the UK

	1974–8	*1985–7*
Deep sea fishing (1959–68)	1 in 360	1 in 1,100
Quarrying	1 in 3,400	1 in 2,600
Coal mining	1 in 4,750	1 in 9,433
Construction	1 in 6,700	1 in 10,900
Agriculture	1 in 9,000	1 in 11,500
Manufacturing	1 in 32,000	1 in 43,000

ICRP recommendations on internal doses

Another major change in dose limitation made by the ICRP in 1977 was the way in which allowance was made for doses to internal organs, mostly as a result of intake of radioactive materials. The ICRP introduced the concept of 'effective dose equivalent', which was the sum of weighted contributions from all irradiated organs to the whole body dose. This was difficult to compare with the existing system in which derived limits of 'maximum permissible body burdens' had been used. These limits, it must be said, had been misunderstood and misapplied since their introduction in 1959 and in *some* cases had resulted in overprotection. However, between 1977–80, with the introduction of a system of assessment of the contribution of organ doses to total risk, the maximum doses recommended for some organs were increased and there was, again, much criticism. The new system was, however, a good first attempt at the summation of internal and external risk by weighting the dose to each major organ as a proportion of the risk from total body irradiation. This produced weighting factors for six major organs and one for all remaining tissues (see Table 3.4).

It was further recommended that a value of W_T of 0.06 was applicable to each of the five organs and tissues of the 'remainder' receiving the highest dose equivalent, and that exposure of other remaining tissues could be neglected. This was later amended in 1978 at the ICRP's Stockholm meeting where it was stated that:

> The Commission wishes to point out that it did not intend the hands, forearms, the feet and ankles, the skin and the lens of the

Table 3.4 ICRP Weighting Factors for Major Body Organs

Organ (T)	*Mortality risk/Sv*	*Weighting factor W_m*
Gonads	40×10^{-4}	0.25
Breast	25×10^{-4}	0.15
Red bone marrow	20×10^{-4}	0.12
Lung	20×10^{-4}	0.12
Thyroid	5×10^{-4}	0.03
Bone surfaces	5×10^{-4}	0.03
Remainder	50×10^{-4}	0.30
Total	165×10^{-4}	1.00

eye to be included in the Remainder. These tissues should therefore be excluded from the computation of W_TH_T. In order to prevent the occurrence of non-stochastic effects the Commission recommends that the relevant dose equivalent limits given in paragraph 103 should apply to these tissues.

Paragraph 103 of ICRP Publication 26 states that 'the Commission believes that non-stochastic effects will be prevented by applying a dose equivalent limit of 0.5 Sv (50 rem) in a year to all tissues except the lens, for which the Commission recommends a limit of 0.3 Sv (30 rem) in a year'. When faced with this set of weighting factors, the logical calculation is to divide the whole body dose limit by each one to deduce a limiting organ dose (see Table 3.5).

The apparent allowable increases in dose over previous recommendations (ICRP 9) are clear from Table 3.5. In fact the apparent absurdity of a stochastic dose limit exceeding the non-stochastic dose limit, which probably represents a threshold, is also clear and had to be avoided in three cases by invoking the non-stochastic limit. The Commission's answer to these criticisms of apparently reduced protection was to argue that as the risk to *all* organs was now considered, most often the whole body risk would be limiting. In addition, they claimed that it was difficult to envisage a situation in which one organ would be irradiated alone. Before 1977, dose limits for single organs, if irradiated alone, still

Table 3.5 ICRP Annual Organ Dose Limits for Radiation Workers (mSv) (Recommended in 1966 and implied in 1977)

Organ	*ICRP 9 (1966)*	*ICRP 26 (1977)*
Whole body	50	50
Gonads	50	200
Breast	150	320
Red bone marrow	50	420
Lung	150	420
Thyroid	300	1670 (500*)
Skin	300	– (500*)
Bone surfaces	150	1670 (500*)
Remainder	150	170

Note * As stated on p. 00 in the extract from ICRP 26, in order to prevent non-stochastic effects, a 500 mSv limit was to be considered limiting.

obviously reflected the notion of a threshold below which damage was either absent or minor. This is evident from the fact that the limit for the whole body with all tissues uniformly exposed was the same as that if only the marrow or only the gonads were exposed (see Table 3.5). This would only be sensible if this whole body dose prevented damage in any body tissue and less sensitive tissues could be ignored if more sensitive ones were protected. It became totally inappropriate when quantifiable carcinogenic risks of different magnitudes were found to apply to different tissues without threshold. Nevertheless, this has been seen by some as a step backwards in radiation protection.

Even more changes were, however, introduced in 1980 when the first part of ICRP Publication 30 was produced. This dealt comprehensively with the derivation of secondary, or derived limits for intake of radionuclides. This publication and subsequent voluminous appendices eventually gave annual limits on intake (ALI) and derived air concentrations (DAC) for all radionuclides used by man.

ICRP recommendations on annual limit on intake

The annual limit on intake (ALI) of a radionuclide is a secondary limit designed to meet the basic limits for occupational exposure recommended by the ICRP and is derived as follows. ALI is the greatest value of the annual intake I which satisfies both of the following inequalities:

$$I \sum_{T} W_T \,(H_{50,T} \text{ per unit intake}) < 0.05\,\text{Sv}$$
$$I \,(H_{50,T} \text{ per unit intake}) < 0.5\,\text{Sv}$$

where I (in Bq) is the annual intake of the specified radionuclide either by ingestion or inhalation; W_T is the weighting factor for tissue T; $H_{50,T}$ per unit intake (i.e. Sv per Bq) is the committed dose equivalent in tissue T from the intake of the unit activity of the radionuclide by the specified route, integrated over 50 years.

In simpler language, ALI is the maximum amount of a radionuclide that may be taken into the body by ingestion or inhalation each year so that neither the stochastic nor non-stochastic dose limit is exceeded.

The derived air concentration (DAC) is another secondary limit used to control intake.

The ICRP defines the DAC as follows:

The derived air concentration (DAC) for any radionuclide is defined as that concentration in air (in Bq per cubic metre) which if breathed for a working year of 2000 hours (50 weeks at 40 hours per week) under conditions of light activity would result in the ALI by inhalation:

$$\begin{aligned} \text{DAC} &= \text{ALI}/(2000 \times 60 \times 0.02) \\ &= \text{ALI}/2.4 \times 10^3\ \text{Bqm}^{-3} \end{aligned}$$

where 0.2 cubic metres is the volume of air breathed per minute

These quantities are, of course, absolutely essential in the practical operation of radiation protection (with radioisotopes) in the workplace.

Unfortunately, ICRP 30 was seen as a direct replacement for, or update of, Publication 2 of 1959 which had listed maximum permissible body burdens and maximum permitted concentrations (MPCs) of radionuclides in air and water. The only point of direct reference can be a comparison of DACs from ICRP 30 and MPCs in air from ICRP 2. There are a total of 260 changes, the derived limits being *increased* for about 170 radionuclides. These changes had a mixed reception, with the opposition being led by Karl Morgan, a member of the ICRP main Commission from 1959 and chairman of its Committee 2 on internal dose for fourteen years, who has published his views a number of times. He welcomed the change in philosophy in ICRP 30, but saw no justification for any limiting derived concentrations being higher than they were in 1959. It is difficult to see these increases as other than a retrograde step in radiation protection. However, the whole philosophy of single organ weighting factors may have to be rethought now that age-specific risks are perceived to be important.

The Ionizing Radiations Regulations

The ICRP publications of 1977 were merely 'recommendations' and therefore not legally enforceable. This situation changed in 1985. After a long gestation period, including extensive periods of consultation and in direct response to an EC directive (80/836/EURATOM 1980), the UK Health and Safety Executive formulated the Ionizing Radiations Regulations (see Table 3.6 for major limits). This Act covers all work with ionizing radiation in the UK and also the exposure of 'other persons', by which was presumably meant members of the public.

Table 3.6 UK Health and Safety Executive Dose Limits (1985)

PART I

Dose Limits for the Whole Body

1 The dose limit for the whole body resulting from exposure to the whole or part of the body, being the sum of the following dose quantities resulting from exposure to ionising radiation, namely the effective dose equivalent from external radiation and the committed effective dose equivalent from that year's intake of radionuclides, shall in any calendar year be:

(a)	for employees aged 18 years or over	50 mSv
(b)	for trainees aged under 18 years	15 mSv
(c)	for any other person	5 mSv

PART II

Dose Limits for Individual Organs and Tissues

2 Without prejudice to Part I of this Schedule, the dose limit for individual organs or tissues, being the sum of the following dose quantities resulting from exposure to ionising radiation, namely the dose equivalent from external radiation, the dose equivalent from contamination and the committed dose equivalent from that year's intake of radionuclides averaged throughout any individual organ or tissue (other than the lens of the eye) or any body extremity or over any area of skin, shall in any calendar year be:

(a)	for employees aged 18 years or over	500 mSv
(b)	for trainees aged under 18 years	150 mSv
(c)	for any other person	50 mSv

3 In assessing the dose quantity to skin whether from contamination or external radiation, the area of skin over which the dose quantity is averaged shall be appropriate to the circumstances but in any event shall not exceed 100 cm^2.

PART III

Dose Limits for the Lens of the Eye

4 The dose limit for the lens of the eye resulting from exposure to ionising radiation, being the average dose equivalent from external and internal

radiation delivered between 2.5 mm and 3.5 mm behind the surface of the eye, shall in any calendar year be:

(a)	for employees aged 18 years or over	150 mSv
(b)	for trainees aged under 18 years	45 mSv
(c)	for any other person	15 mSv

PART IV

Dose Limit for the Abdomen of a Woman of Reproductive Capacity

5 The dose limit for the abdomen of a woman of reproductive capacity who is at work, being the dose equivalent from external radiation resulting from exposure to ionising radiation averaged throughout the abdomen, shall be 13 mSv in any consecutive three month interval.

PART V

Dose Limit for the Abdomen of a Pregnant Woman

6 The dose limit for the abdomen of a pregnant woman who is at work, being the dose equivalent from external radiation resulting from exposure to ionising radiation averaged throughout the abdomen, shall be 10 mSv during the declared term of pregnancy.

Thc interested reader could obtain a copy of these regulations from HMSO, together with various Codes of Practice that have been issued subsequently. These codes are themselves fairly complex and do not always shed more light than the original regulations, but in many respects the exercise has been a laudable attempt to incorporate radiation protection recommendations into legislation. In general, these regulations endorse ICRP's recommendations, which is in itself a contrast to the current recommendations of the UK's National Radiological Protection Board (NRPB). They give complex details of control of radiation in the workplace, classification of work areas, the criteria for designation of workers, and requirements for monitoring and medical surveillance. They also require total (i.e. the sum of internal and external) dose to be recorded for radiation workers, which has caused some problems in the nuclear industry. However, the regulations adopt a 500 mSv dose equivalent limit for all individual organs. This could have the effect of allowing the annual *effective* whole body dose limit to be exceeded if the ICRP weighting factors are used. For example, for the lung the formula is $0.12 \times 500 = 60$ mSv, which is 10 mSv over the dose limit.

However, it is noteworthy that the ICRP organ weighting factors do not appear in the Regulations. This aspect again seems to be sacrificing safety for simplicity. Nevertheless the Regulations do specify an annual investigation level of 15 mSv. This is a good step as long as the investigation is thorough and more than just a record of the event.

The body giving the government advice on these matters, the NRPB, is now adopting a stance some way from the basic dose limits of the Regulations. It remains to be seen whether its recommendations will be endorsed by legislation or whether the government and its advisers will remain polarized.

The Biological Basis of Dose Limits

Assessment of risk is central to the concept of dose limitation and control. For application to situations of low-level chronic exposure, this assessment must include not only the probability of cancer induction (not just cancer mortality) but also a component for genetic damage, i.e. in ICRP parlance, a summation of the stochastic effects. To set a dose limit, this total risk must then be compared with risks involved in other human activities in order to judge its acceptability or even tolerability. Dose limits can be set in this way using different standards of judgement, for radiation workers and for members of the general public, the latter limits being used to control discharges and radioactive waste disposal as well as for establishing generalized derived limits (GDLs). The two parts of this process, the derivation of a risk rate per unit dose and judgement over acceptability of risk, will be considered separately, with indications of where controversy exists.

Risk Rates for Cancer Induction

From what has been stated above, it will be clear that information on the risk of cancer mortality comes from epidemiological studies. These are periodically reviewed by international committees and organizations and best estimates (or ranges) of risk rates are published. The most active in this field is the United Nations Scientific Committee on the Effects of Atomic Radiation (UNSCEAR) which has published quite large reviews about every four years. Its latest report, published in 1988, is worth close scrutiny.

Another committee of some repute is that jointly set up in the USA by the National Academy of Sciences and the National Institutes of Health. This Committee, called the Biological Effects of Ionizing Radiation Committee (BEIR), first reported in 1972 (BEIR I) and then in 1980 (BEIR III). BEIR IV, which covered the effects of alpha emitters, was published in 1988. BEIR V, which is to cover cancer risk rates for low-LET radiation, is eagerly awaited. The ICRP committees also review the available data but the ICRP has tended to rely heavily on UNSCEAR for guidance. A relative newcomer to this field of risk rate assessment is the National Radiological Protection Board (NRPB) whose role of adviser to the UK government has usually been fulfilled by endorsing ICRP recommendations. This seems to have changed now and, as will be seen later, the NRPB has upstaged ICRP by recommending the use of risk rates that are up to six times those previously accepted.

However, this apparent disparity between reviewing bodies is not new. Table 3.7 shows the wide spread of risk rates for all-cancer mortality that have been calculated over the last decade or so, mostly from the same basic data. Let us therefore now consider uncertainties in the way in which latest estimates of radiation risk have been derived and the implications of these risk estimates in terms of dose limits.

Table 3.7 Lifetime Population Cancer Risk Rates

Source	*Deaths/10,000 people/Sv*
BEIR I (1972)	117–621
UNSCEAR (1977)	250* (range 75–175 low dose)
ICRP (1977)	125
BEIR III (1980)	167–501
NRPB (1987)	300
UNSCEAR (1988)	700–1,100* (relative risk)
	400–500* (absolute risk)
NRPB (1988)	1,290* relative risk
	450 low dose rate
	368* absolute risk
	131 low dose rate

Note *High dose rate

Projecting risks over a lifetime

There are very few epidemiological studies that have followed an entire population to the end of life. At present, the only such study is on early US radiologists. However, the dosimetry in this study is somewhat uncertain, with doses of between 100 and 2000 rad (1–20 Gy) lifetime dose having been reported. Eventually it is hoped that a study on UK radiologists will be completed but it is unlikely that the dosimetry will be any more certain.

The Life Span Study (LSS) cohort of Japanese bomb survivors has now been followed for forty years and so another 40 years will probably be necessary for determination of their total risk. Two-thirds of this group were still alive in 1988, so some means of projecting the risk beyond the limited period of follow-up is required.

For leukaemia this does not appear to present a great problem. Data from Japan and some of the other medical studies have shown a similar temporal pattern, with excess rates of leukaemia appearing less than three years after exposure, a rise to a peak at about five to seven years, and then a decline. However, the latest results still show a slight excess forty years after exposure. Follow-up of the Japanese did not begin until 1950, when the ABCC was formed. Leukaemias that may have occurred in this first five-year period have been estimated by comparison with the ankylosing spondylitics study and by back extrapolation of the observed dose–response expression. There are also good data to indicate that the bone cancer risk pattern is close to that of leukaemia. However, for other (solid) cancers, the risk does not appear to be tailing off with time.

To account for this temporal expression of risk a number of models have been developed to fit the observational data. The two models in use for at least the last decade are the absolute (additive) and relative (multiplicative) risk models. The pattern of risk that these two models describe is shown diagrammatically in Figure 3.2.

With the absolute risk model, the absolute excess risk remains constant after a latent period has been exceeded. With the relative risk model, the risk relative to the baseline (natural or spontaneous) risk remains constant with time, again after a latent period has been exceeded. Since the baseline rates for solid tumours increases rapidly with age (see Figure 3.3), the constant

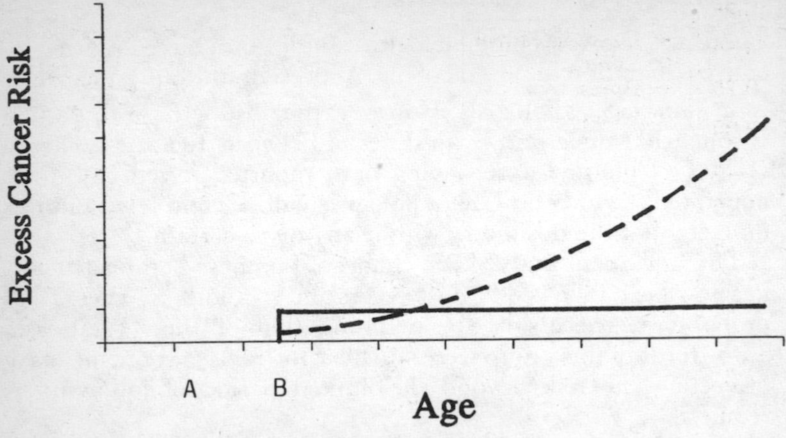

Figure 3.2 Absolute and Relative Risk Models.

Note: After irradiation at age 'a' effects are seen following a latent period at age 'b'. The absolute risk model predicts an initial absolute increase in the number of cancers which then remains constant. The relative risk model predicts an excess of cancers which is a constant multiple of the baseline (or spontaneous) incidence. The number of cancers therefore increases with age as shown by the dashed curve above.

relative risk model yields a much larger (two to five times) predicted number of cancers than the absolute risk model.

These models are used to fit the observed data; they say nothing about the biology of the induction of cancer. However, the relative risk model provides a good fit if radiation's effect is to initiate carcinogenesis, whereas the absolute model fits well if radiation has a promoting effect. In practice, radiation probably has both effects so that hybrid extrapolation models would possibly fit best of all.

Most of the data now coming from a variety of studies – Japanese bomb survivors, studies of radiation-induced breast cancers in women and skin cancer in the children irradiated for tinea capitis – show a much better fit to a constant relative risk model. The only groups of the Japanese who appear to exhibit a lessening of (relative) risk with age are those who were under 10 at the time of the bombings in 1945, but the data are really too few to be certain of this trend. There are, however, a few other studies in which the relative risk has not remained constant whereas the absolute risk has, e.g. studies of thyroid cancer after irradiation in

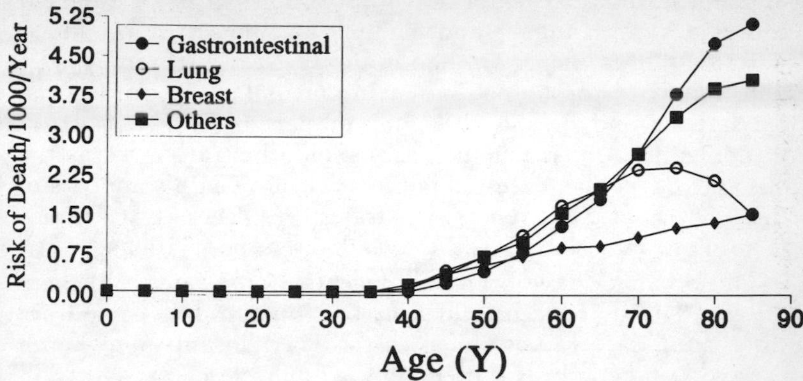

Figure 3.3 Baseline Risk of Death from Four Types of Cancer (Gastrointestinal, Lung, Breast, and Other) as a Function of Age.

childhood. The ankylosing spondylitics study is, as has been previously described, anomalous in that all cancers fit an absolute model somewhat similar to leukaemia, with very few cancers appearing twenty to twenty-five years after exposure.

The ICRP used an absolute risk projection model for endorsing its recommendations in 1977 but by 1980 it was beginning to acknowledge that relative risk models might provide better risk estimates. The BEIR III committee in 1980 gave risk rates for both models. By 1988, the UNSCEAR was stating that 'recent findings in Japan suggest that the relative risk projection model is the more likely, at least for some of the more common cancer types'. This must be seen as an important statement because it implies a significant underestimate of cancer mortality hitherto. For instance, the NRPB has calculated that the predicted lifetime risk will be about four times the risk that could be estimated from the experience in other populations, e.g. the Japanese, to date.

Transferring risks between populations

One of the problems of employing a relative risk model is in the transference of risks from one population to another. This is because the baseline rates for different cancers are significantly different in different countries. For example, rates for stomach cancer are much higher and for lung and breast cancer much lower (six times) in Japan than in the West. Apart from genetic differences, it is probable that environmental and social factors, e.g. smoking and diet, have a significant part to play in explaining this. The problem is whether risks should be transferred as the absolute or relative excess. Figure 3.4 (from NRPB's analysis of the 1988 UNSCEAR report) illustrates the effect of transference of absolute and relative risks to two different populations.

This seems to have been only recently appreciated and there are few data that are helpful. The best human data come from three studies of radiogenic breast cancer: in Japanese bomb survivors, tuberculosis patients given multiple chest X-rays in New England, and postpartum mastitis patients given X-ray therapy in New York. It was found that the age-specific absolute excess risk was similar for all three populations. As the baseline breast cancer rates are much higher in the USA than in Japan, the relative risk for the Americans was lower than for the Japanese.

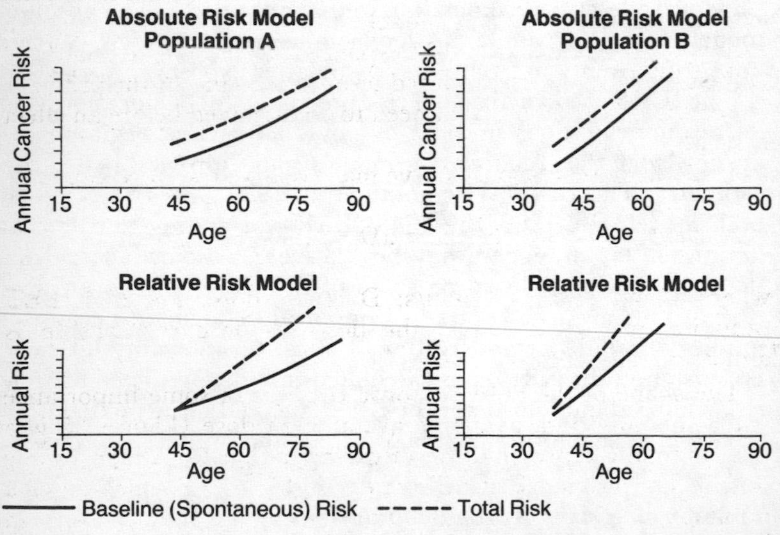

Figure 3.4 Radiation-induced Cancer Incidence Under Absolute and Relative Risk Models for Two Populations with Different Spontaneous Cancer Incidences.

However, the only sound animal data suggest that transference of relative risk is correct.

More generally, in the Japanese LSS cohort the absolute risk, apart from leukaemia, tends to vary across cancer types in much the same way as the baseline risk. Thus the evidence tends to favour a model in which radiation acts multiplicatively rather than additively on the baseline risk. Relative risk coefficients have therefore more often been used for transferring risks from Japanese to UK populations (except for leukaemia). However, it is clear that for some tumours the process of transference of risk can introduce differences of a factor of three. Another uncertainty.

Extrapolation to low dose and low dose rates

DOSE

There has been considerable debate over the shape of the dose–response curve for radiation-induced cancer. Conventional radiobiological wisdom suggests a linear relationship between dose and probability of cancer. This has been supported by a wide range of empirical low dose data on cells in culture and some animal experiments. However, at higher doses for low-LET radiation (X-rays or gamma rays) the response may be linear quadratic with an additional dose-squared term to account for the presence of two biological targets that need to be damaged before an effect is seen.

Thus the dose–response curve may well be described by:

$$\text{Effect} = aD + bD^2$$

where a and b are constants; D is the dose. For high LET radiation (e.g. alpha rays) the dose–response relationship is almost certainly linear.

The shape of the dose–response curve is of some importance when the only data available are at high dose (Figure 3.5 is a diagrammatic representation of extrapolation from high to low doses). The problem is also illustrated in Figure 3.6 which shows a number of curves fitted to the leukaemia data shown in Figure 2.2. All fit reasonably well but imply somewhat different risks at low doses. Although animal data have shown linear quadratic and other dose–response curves, the best fit to the

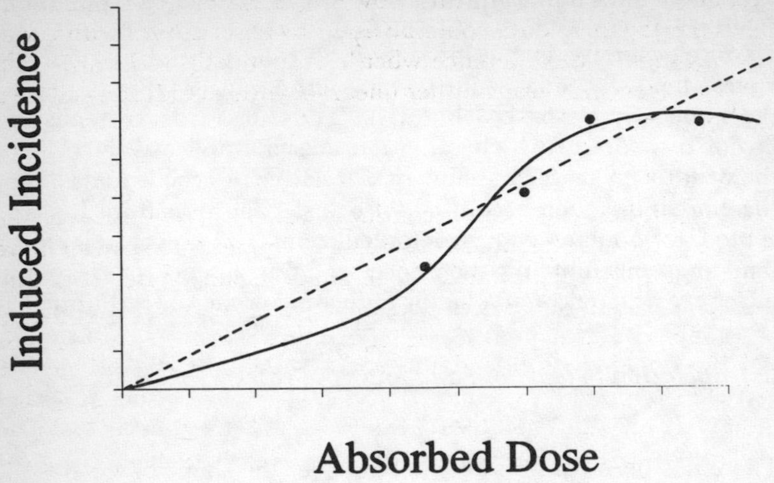

Figure 3.5 Extrapolation from High to Low Doses.

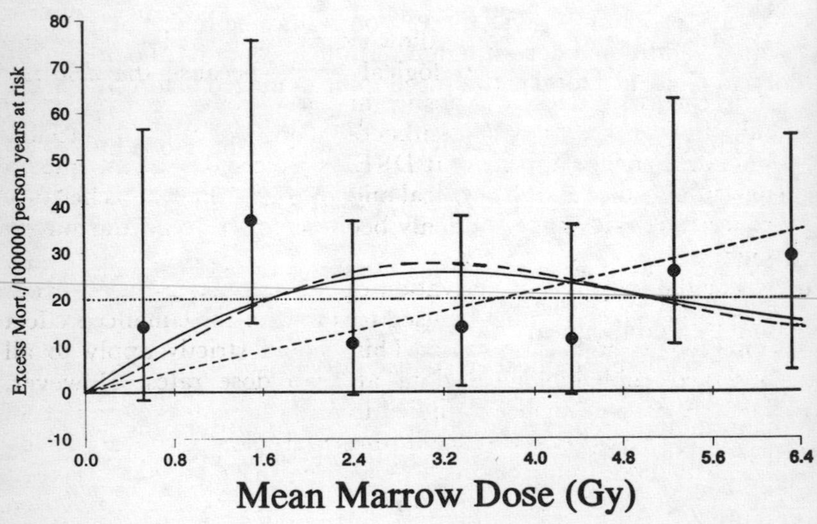

Figure 3.6 Various Curves Fitted to Data of Leukaemia Deaths Following Irradiation for Ankylising Spondylitis

Japanese data now is linear. However, it should be pointed out that the Japanese data could be fitted by other curves with equal justification. The difference when extrapolating backwards to a dose of a few mSv using either linear or linear quadratic can be a factor of two to three.

DOSE RATE

Although there has been a continuing and long debate over the shape of the dose–response curve, there promises to be even more discussion of the allowance to be made for dose rate. It is well established that repair occurs in most biological systems following radiation damage and it seems logical that this will include repair of the relatively subtle damage that might be the prelude to carcinogenesis. This concept can explain the dose–response relationship which includes a dose squared term (a linear quadratic) but is difficult to reconcile with a linear relationship. A linear dose–response implies that the effect per unit dose is the same at all doses. However, the latest data from the A-bomb survivors could be fitted by a linear curve (best) or possibly a linear quadratic.

Nevertheless, as soon as the latest cancer risk rates were calculated for the Japanese, a dose rate effectiveness factor (DREF) was introduced to allow extrapolation to low dose rate. This is a reasonable and logical action because the *dose rate* difference between the exposure in Japan and that of a typical radiation worker, or even member of the public, is a factor of a few million. It might appear that DREF's have suddenly assumed a new importance and the cynical might suggest that this is because cancer risk rates have suddenly been perceived to be alarmingly high!

For the purposes of derivation of risk rate for low-level exposures, these factors must be used to allow for the enhanced effect observed at high dose rates. This should strictly apply to all observed epidemiological data at high dose rates. However, DREF's have never been applied to any study other than that of the Japanese bomb survivors for these purposes. This is confusing because data from medical exposures often at similar high instantaneous dose rates, but without a DREF applied, have been cited as providing evidence of low dose risk rates.

However, the concept of a reduced effect with lower dose rate is not new as a DREF of 2 was used in 1977 by ICRP in deriving its cancer mortality risk rate of 1.25 per cent/Sv (1.25 in 100 people per Sv) from the UNSCEAR figure of 2.5 per cent/Sv. The Table 3.8 shows how DREF values have varied in the years since.

Most values of DREF have been derived from animal radiation carcinogenesis data which have often been found wanting for extrapolation to humans because of the use of inappropriately high doses and dose rates, and application to tumours that do not occur in man. There is also good evidence of a large range in DREF between tumour types in animals, most of which are bred to exhibit a certain cancer. The first review of these animal data was published in 1980 in a report of the US National Council for Radiological Protection and Measurements (NCRP) and figures since have generally been derivatives of those data with some dubious back-up from human data.

It has often been stated that, for purely statistical reasons, human epidemiology yields data on cancer incidence versus dose that makes any distinction between various dose–response relationships very difficult, and differences due to dose rate effectiveness impossible to detect. However, this has not deterred attempts to use human data to demonstrate DREFs different from 1. A few examples will serve to illustrate the relative futility of this exercise.

The first study concerns the induction of lung cancer in uranium miners. The arguments are fairly convoluted but, briefly, depend on risk rates for lung cancer in miners after protracted exposure to radon compared to risk rates derived from the Japanese bomb survivors. When the Japanese data are divided by a DREF of between 2 and 5, the risk rates are similar to those found in the miners. Such arguments depend on the estimation of lung dose and the Relative Biological Effectiveness (RBE) chosen for the alpha radiation from radon and some of its daughters. Neither of these two parameters are known with any

Table 3.8 DREFs Used Since 1977

UNSCEAR/ICRP 1977	2
NCRP 1980	2–10
UNSCEAR 1986	'up to 5'
UNSCEAR 1988	2–10
NRPB 1988	3

certainty and, together with the confusing effect of smoking, make this estimate of DREF tenuous, at least.

Another study of health effects that has been used to substantiate a figure for DREF is that on US radiologists. DREFs of about 5 for life shortening and 2.7 for leukaemia have been derived from this group. These workers no doubt received radiation exposure at a fairly low dose rate over long periods but estimation of this dose has proved difficult. For instance, the NCRP quotes a twenty-fold variation in published estimates, with 6 Gy as a mean. There seems so much uncertainty in the estimate that any DREF derived must be speculative. These data were viewed by NCRP as being fairly weak.

The NCRP also quoted results from high dose rate studies (these included both medically exposed groups and the Japanese bomb survivors) and low dose rate studies (the US radiologists were one) and attempted to compare them; their conclusion was that: 'As with the evaluation of dose effect, we believe that inter-study comparison cannot be interpreted relative to a possible dose-rate effect'.

Lastly, some emphasis is currently being placed on follow-up studies of people treated with external (high dose rate) and internal (protracted and therefore lower dose rate) radiation that has produced thyroid cancer. These data were again reviewed in 1980 by NCRP. At that time the study suffered from a lack of numbers of cases and was therefore dismissed. However, in a later review of the same cohort by UNSCEAR in 1986 a DREF of 2 was estimated for adults and 4 for children based on a comparison of external irradiation and iodine-131 (protracted) internal radiation. UNSCEAR believed that the differences in effect were explicable by the distribution of iodine in the thyroid. However, it is clear from a 1985 NCRP review that iodine-131 did not appear to induce *any* thyroid cancers. For instance, the summary begins:

> iodine-131 appears less carcinogenic in people on a rad-for-rad basis than external radiation. How much less is yet to be determined; in fact available human data on low dose iodine-131 exposures have not shown iodine-131 to be carcinogenic in the human thyroid.

This statement is in no way contradicted by two later papers (1988–9) on a longer follow-up of people exposed to iodine-131 and external radiation in Sweden. The conclusions of these papers were that:

> Overall these data provide little proof that iodine-131 is carcinogenic in humans and support the notion that the carcinogenic potential of internal iodine-131 beta particles might be as low as four times less than external X-rays or gamma rays.

Interestingly, another paper in 1982 had concluded, on the basis of *animal* experiments (when iodine-131 *did* induce some thyroid tumours), that there was no difference in the effectiveness of iodine-131 and X-rays! These animal data seem to have been ignored.

Clearly, although a DREF seems applicable here, there is no sound evidence for quantification unless experimental data are presented of a dose rate effect for the same radiations, and selective use of data is to be questioned.

UNSCEAR, in its 1986 report, fully considered the effect of dose rate, and produced the following guide:

> The overestimation of the probability of effects at about 10 mGy from single-dose data at 1–2 Gy (acutely delivered) by linear (as opposed to linear-quadratic) extrapolation would vary from 1.5 to 3.0 for an assumed reasonable set of parameters.
>
> Similar over-estimation would apply to the effects of low-LET radiation delivered at low dose rates irrespective of the total doses below the level when cell killing becomes important. This would result from the disappearance of the quadratic component of the dose–response curves owing to repair of sublesions.

On the basis of these somewhat confusing statements about theoretical models and also from animal data, UNSCEAR 1988 have re-advised that DREFs should be in the range 2–10. The NRPB, in its latest report, has recommended the use of 3, this being the median of the above range (they use a value of 2 for breast cancer).

Clearly, the actual figure to be used is a matter of some importance and the evidence is to a certain extent contradictory. There has been a tendency for selective use and reporting of data, and this further confuses the situation. In the absence of any sound guidance, it seems it would be prudent to err on the conservative (safety) side. Thus, I recommend the use of a figure of 2 applicable to all situations and therefore prudent and conservative. Because this one figure would be used for all cancers to simplify the dose limitation system, it must be at the bottom end of the range.

Perhaps a final word on the subject of DREF should be left to one of the more prestigious committees, the NCRP, which in a 1989 report stated that: 'Great uncertainty attaches to the proper numerical value [of DREF] suitable for human carcinogenesis, and to whether, indeed, there should be a single value, owing to the dearth of relevant information.'

Risk Rates for Fatal Cancer Suggested by NRPB

As soon as the UNSCEAR report of 1988 was published, the NRPB set about the task of applying the data to UK populations and deriving risk rates. In doing this it drew largely on the models recommended by UNSCEAR for application to the Japanese bomb survivor epidemiology, but other studies were included where appropriate. Its results will be discussed here because they are the most appropriate at present to apply to a UK population.

Using a relative risk extrapolation model, the NRPB estimated the risk of all fatal cancers to the UK population (all ages, both sexes) as 12.9 per 100 people per Gy derived from high dose rate data. This figure is shown in Table 3.9. This also shows the result of applying the NRPB's DREF of 3 (2 for breast cancer) and also the risk to date, i.e. without the use of a lifetime extrapolation model. It is clear from Table 3.9 that an upper bound for fatal cancer can be taken as about 13 in 100 per Gy but a more likely estimate is about 4.5 in 100 per Gy. This figure is about 3.6 times the risk estimated in 1977 by ICRP, which is the basis for dose limitation in the UK at present. It is worth noting that for some cancers risks estimated now are similar to 1977 estimates (bone, liver, and leukaemia), whereas for some (breast, colon, stomach) they are very different.

For application to a working population (aged 20–64) the NRPB cancer mortality risk rates are a little different. They are shown in Table 3.10. They are lower because children have been

Table 3.9 Estimated Excess Cancer Risks in a UK Population (All Ages Both Sexes) Associated with Exposure to Low LET Radiation

	Fatal cancer risks, 10^{-2} Gy^{-1} (low LET)				
	Lifetime projection		*Risk observed to date*		*ICRP*
Cancer type	*HDR*	*LDR*	*HDR*	*LDR*	*1977 (adults)*
Leukaemia	.84	.28	.84	.28	.2
Breast	1.1	.55	.42	.21	.25
Lung	3.5	1.2	1.15	.38	.2
Thyroid	.075	.025	.055	.02	.05
Bone	.15	.05	.15	.05	.05
Liver	.45	.15	.23	.08	.1
LLI/colon	1.1	.37	.38	.13	.1
Stomach	.73	.24	.14	.05	.1
Remainder	5.0	1.63	.565	.20	.2
Total	12.9	4.5	3.93	1.4	1.25

Note HDR is 'high dose rate'. LDR is 'low dose rate', i.e. after application of a DREF.

excluded, but again the overall result appropriate to low dose rate exposures is still about 3.4 times the risk used by ICRP. It is also a little higher (3.4/3) than the risk rate used by the NRPB in its report GS9 issued in 1987. This recommended an average dose limitation of 15 mSv (just under one-third of the existing dose limit).

From what has been said earlier it will be clear that there must be large uncertainties attached to such estimates. For the purposes of protection it is to be hoped that this results in overestimates of risk, but this is not guaranteed. The uncertainties have been covered earlier in this chapter but it is worth emphasizing two of these:

(a) DREF is not known with any precision and varies with tumour site. To be conservative a DREF at the lower end of any range should be chosen if it is to apply to all cancers.

(b) The transference of data, largely from a Japanese population, to a Western population presents difficulties and there are several suggestions as to how this should be done. For instance the UK's Central Electricity Generating Board used a transfer methodology that results in an overall non-leukaemia cancer risk rate which is a factor of three

Table 3.10 Estimated Excess Cancer Risks in a UK Working Population (Both Sexes) Exposed to Low LET Radiation at Ages 20–64 Years

	Fatal cancer risks, 10^{-2} Gy^{-1} (low LET)				
	Lifetime projection		*Risk observed to date*		*ICRP*
Cancer type	*HDR*	*LDR*	*HDR*	*LDR*	*1977*
Leukaemia	1.01	.34	1.01	.34	.2
Breast	.79	.40	.51	.26	.25
Lung	2.09	.70	1.64	.55	.2
Thyroid	.05	.017	.03	.01	.05
Bone	.15	.05	.15	.05	.05
Liver	.45	.15	.23	.08	.1
LLI/colon	1.21	.40	.56	.19	.1
Stomach	.59	.20	.19	.06	.1
Remainder	3.33	1.10	.88	.28	.2
Total	9.67	3.4	5.20	1.82	1.25

Note HDR is 'high dose rate'. LDR is 'low dose rate', i.e. after application of a DREF.

lower than those derived from the RERF Japanese data. Clearly the results are sensitive to this parameter.

How realistic are these figures as a measure of total risk to people exposed to low doses? The answer to this is 'not very'. To start with, this value only covers fatal cancer. There is now a belief that *total* cancer incidence is important, i.e. the probability of contracting cancer rather than only the risk of dying of it. These data are notoriously difficult to gather reliably, but with the quality of cancer treatment varying markedly they are obviously important. There are about 50 per cent more non-fatal than fatal cancers and therefore risk rate figures should be multiplied by 1.5 to apply strictly to total cancer risk. However, in its evidence to the Hinkley Point Inquiry in 1989 the NRPB chose to make a value judgement over the importance of non-fatal cancer. It used a 'weighting factor' to allow for the contribution of non-fatal cancers to total risk. For cancers such as skin cancers, which are rarely fatal and of little significance in terms of impaired life, this is reasonable, but for other cancers this must result in an underestimate of total health detriment.

In addition *total* risk must include a component for serious genetic damage in all generations. This has been estimated by

NRPB from UNSCEAR data as 0.8 cases in 100 people per Gy which, as stated before (Chapter 1), covers the extrapolated estimates. From these data the NRPB estimated a total risk to a UK population of 5.6 per 100 per Gy, and to a working population of 4.5 per 100 per Gy. These figures represent risks that are about five times those used by ICRP in 1977.

Dose Limits

The setting of a dose limit from data on risk must include some consideration of what risk is 'tolerable' or 'acceptable'. This has traditionally been done by comparing radiation risks with other risks that are considered acceptable, at least as a trade-off for some benefit. In 1977 the ICRP equated risk at the dose limit with work in a 'fairly risky' industry and risk at the average dose (about a tenth of the dose limit) with work in traditionally 'safe' industries. Two aspects of this comparison need further consideration. First, it is questionable whether fatal accident rates in industry (see Table 3.3) are a suitable comparison with delayed effects such as cancer. Secondly, accident rates in most industries have improved since 1977 whereas risk rates from radiation exposure are considered to be much higher than was previously believed. It is therefore questionable whether the industrial comparison would appeal to the nuclear industry any more.

The whole question of risk acceptance has been the subject of two recent reports by the Health and Safety Executive (HSE) and the Royal Society, respectively. Their conclusions differ a little.

The Royal Society report argues that an unacceptable *imposed* risk will be one at a level between 1 in 10,000 and 1 in 100,000 per year. The HSE report, by contrast, states that: 'Broadly a risk of death of 1 in 1000 per annum is the most that is ordinarily accepted under modern conditions for workers in UK and it seems reasonable to adopt it as the dividing line between what is just tolerable and what is intolerable'. It is interesting that in 1977 the ICRP considered a risk of 1 in 10,000 to be the maximum that a worker would consider acceptable! The Royal Society seemed to be at variance with this although it could be argued that its acceptable band for voluntary risks would just include the HSE limit. The HSE arrived at a tolerable limit for the public of 1 in 10,000 by reducing worker limits by 10, whereas the Royal Society began with acceptable risks being in the range of 1 in 1,000,000 to 1 in 100,000 and argued the unacceptable figure up towards 1 in 10,000 per year. This seems to indicate a risk of about 1 in 10,000 per year verging on the unacceptable but presumably the totally unacceptable band starting at 1 in 1000. The NRPB has used HSEs limits of tolerable risk to assess risks at various annual dose levels (see Table 3.11).

Implications of Increases in Risk Factors for Dose Limitation

It is clear that a degree of value judgement is required to derive a dose limit from a figure for total risk. If the HSE's views are accepted, a risk of 4.5 per cent/Sv would result in an annual dose limit for radiation workers of about 22 mSv or, if the ICRP risk of 1 in 2000 per year at the dose limit is used, 11 mSv per year. If the Royal Society's tolerable risk limit of 1 in 100,000 per year is used for members of the public, the dose limit for them would be 0.2 mSv/year.

Bearing in mind that over 1000 radiation workers in the UK receive more than 15 mSv/year and target doses for exposures of critical groups of the public have been 0.5 mSv/year, the limiting doses calculated above will present a considerable problem to the nuclear industry and others. It must also be remembered that, if a

Table 3.11 Annual Risk Levels Corresponding to Continuous Exposure at Certain Levels of Dose

Annual dose	*Effective risk* *(10^{-2} Sv^{-1})*	*Annual risk*
Occupational	4.5	
50 mSv – ICRP and UK limit		2×10^{-3}
15 mSv – NRPB interim guidance		7×10^{-4}
HSE limit of tolerable risk		10^{-3}
Public	5.5	
1 mSv – ICRP principal limit		6×10^{-5}
0.5 mSv – NRPB interim guidance		3×10^{-5}
HSE limit of tolerable risk		10^{-4}

DREF of 2 is used and full weight is given to non-fatal cancers, risk rates about ten times the 1977 ICRP risks obtain, i.e. about 10.5 per cent/Sv. Risk rates of this magnitude would mean, if translated into dose limits as above, very far-reaching changes in work practices and would necessitate policies of near-zero discharges. Although the nuclear industry pays lip service to this ideal, it is not yet ready to implement it.

Are we therefore expecting more than is reasonable? Chapter 10 considers whether we are over- or under-protected.

Addendum: The BEIR V Report

This report was published after the main text of Chapter 3 had been written. It is important that a comment on this should be included.

At the end of 1989, the fifth report of the US Committee on the Biological Effects of Ionizing Radiations (BEIR V) was published. This committee reviewed again all available data on the long-term risk of exposure to radiation and produced lifetime risk estimates for external and internal radiation.

Overall, this latest review contains data on genetic risks and cancer risks, foetal effects and epidemiology with no dramatic surprises. However, their approach to cancer risk was somewhat different from UNSCEAR and therefore warrants some consideration. For their detabase, they used the major epidemiological studies including the survivors at Hirohsima and Nagasaki and the various groups exposed to radiation for medical purposes.

The found that the leukaemia mortality data were best fitted by a linear quadratic dose-response model (that is, a curve of the form $aD = bD^2$, where D is the dose and a and b are constants), whereas the mortality datafor all other cancers were best fitted by linear dose-response models. For extrapolation to a lifetime experience, the committee used relative risk models although, unlike UNSCEAR, with modifying factors. For instance, only leukaemia mortality was fitted by a *constant* relative risk model (after a two-year latency period), other cancers (grouped as respiratory, digestive and other conacers) were fitted by relative risk models (after a ten-year latent period) with a time modifying factor. This sort of model had a dramatic effect in the case of respiratory cancers where the relative risk decreased by a factor 5 over the ten to thirty years following exposure. However, the committee also noted that other models, even additive or absolute, gave reasonable fits to the available data.

The committee acknowledged that as all the data were from situations of high dose rate (including the medical radiation), they felt that some use should be made of a dose rate effectiveness factor (DREF) to decrease the effects predicated for low dose rate applications. However, they noted that in practice the use of a linear quadratic dose-response model for extrapolating from high to low dose (for leukaemia) implied a DREF of about 2. Also, confusingly, the committee pointed out that for all other cancers the quadratic component was nearly zero (i.e. the dose response was essentially linear) and consequently the estimated DREFs were near unity. Thus, no DREFs were deemed necessary to apply to any cancer risk estimates. In fact, the committee seemed a lot less than equivocal about this; for example, they say that 'it may be desirable to reduce the estimates derived here by a dose rate effectivness factor (DREF) of about 2 for application to populations of persons exposed to small doeses at low dose rates'. However, they then go on to point out that high energy gamma rays (the source of exposure at Hiroshima and Nagasaki) are reported to be only half as effective as mid energy X-rays. The suggestion (no more) is that these two effects would cancel each other out!

Two important aspects of the review are shown in Figure 3.7 – the age and sex specificity of cancer mortality risks. Of great interest are the shapes of these curves; for instance, it can be seen that, although the radiation sensitivity at ages below about 25 years of age is high, the sensitivity above that age is more or less constant (except for respiratory cancers). These result patterns are considerably different from those in the last related BEIR report (BEIR III, 1980).

The estimated lifetime risks are shown in Table 3.12 for three different possible exposure scenarios. These may be compared with those in Table 3.9. Taken straight from these two tables, it would appear that the BEIR V estimates of total cancer mortality are higher than the risks derived by the NRPB from the UNSCEAR 1988 report by a factor 1.75. However, the marked difference in model approach, e.g. the use of a linear dose-response model for all cancers except leukaemia, the use of relative risk lifetime extrapolation models for all cancers and the absence of a dose rate effectiveness factor, make the similarity of these two risk estimates quite remarkable. The importance of the BEIR V review lies in its chronological position just before the next pontification of ICRP on dose limits (due in late 1990, early 1991). This review again underlines the perception of the increased risk of radiation exposure and the importance of protecting workers and the public with lower limits and/or new radiation protection criteria.

Table 3.12 Cancer Mortality; Lifetime Risks per 100,000 exposed people (BEIR V)

	Male			*Female*		
	Non-leuk	*Leuk*	*total*	*Non-leuk*	*Leuk*	*total*
Single exposure to 100 mSv	660	110	770	730	80	810
Range limits*	420–1040	50–280	540–1240	550–1020	30–190	630–1160
Continuous life-time exposure to 1mSv/yr	450	70	520	540	60	600
Range limits*	320–830	20–260	410–980	430–800	20–200	500–930
Continuous exposure at 10mSv/yr from age 18–65	2480	400	2880	2760	310	3070
Range limits*	1670–4560	130–1160	2150–5460	2120–4190	110–910	2510–4586

Notes

* Range limits are given such that there is only a 10 per cent chance of an estimate of risk being outside the range.

(1) Estimates for leukaemia contain an implicit dose rate rduction factor because of the use of a linear quadratic dose-response model.

(2) No dose rate reduction factor has been applied to the risk estimates for solid cancers.

(3) All risk estimates are age weighted averages.

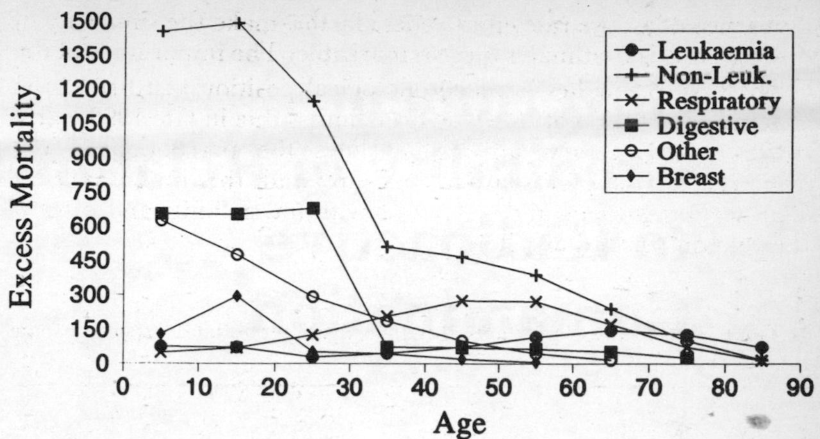

Figure 3.7 Cancer Excess Mortality by Age at Exposure for 100,000 Persons of Each Age Exposed to 0.1 Sv (female, above; male, below).

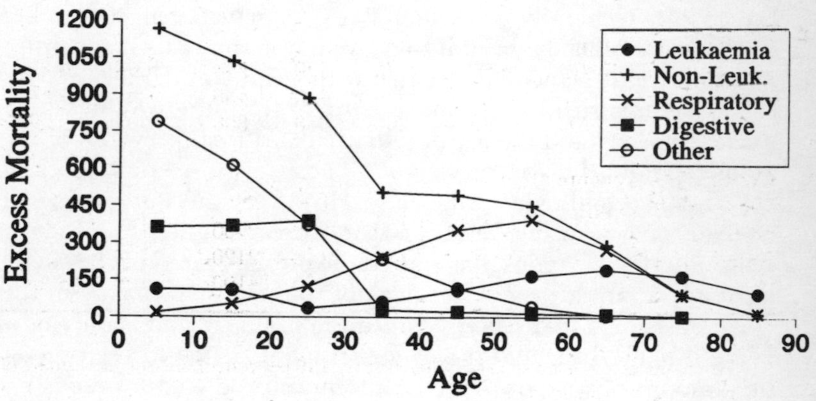

4 Government Credibility in Relation to Radioactive Contamination

The complexity of the whole subject of radiation hazards means that to a large extent the public relies on experts for information and reassurance. Authoritative information has not been easy to come by and this has bred an almost irrational fear in the minds of lay people. Generally the 'experts' are, or have been seen to be, government officials (even if they are not actually civil servants) and their reassurances are treated with suspicion. Unfortunately this basic mistrust stems almost entirely from the way in which nuclear power (and the use of radiation as a tool) grew out of the atomic bomb. This is regrettable, because the medical uses of radiation undoubtedly came first. However, the burgeoning of nuclear power, using reactors that were designed to produce plutonium for weapons, has soured the public opinion. The single most important aspect of this negative public image has been the cloak of secrecy that has enveloped the industry and is only now being lifted. Given that it was government policy to keep secret the amount of weapons-grade plutonium produced by the early nuclear power industry, it may seem reasonable that there was an information black-out. However, it was unfortunate that this had to extend to levels of waste discharged and subsequently to the details of the fire at Windscale in 1957. Officials of the United Kingdom Atomic Energy Authority (UKAEA) were anxious to protect the embryonic nuclear power industry of 1957 by not admitting to the possible health effects of iodine-131 and polonium emitted at the time. However, when full details were released, suspicions were confirmed.

The Windscale Fire

On 8 October 1957, a physicist at Windscale started a series of operations designed to release energy stored in the graphite moderator of the No. 1 reactor. This energy, known as Wigner energy, was periodically and routinely released by a cycle of high and low power levels. On this occasion he noticed that the temperature of the reactor was falling without the Wigner release having been completed. He withdrew more control rods to increase the power and therefore the temperature. Unfortunately, unknown to him, the temperature recording thermocouples were not in the hottest part of the core and this part was already too hot. At 11.05 am the increase of power caused one or more of the fuel rods to ignite. The instrumentation in the reactor was inadequate and the fault was not detected until 5.40 am on 10 October. By then the fire had been raging for more than 42 hours.

The first indication that there was a fire came from monitors in the filters on top of the cooling air discharge stack. Ironically, these filters, still one of the characteristic features of the site, had been added as an afterthought on the order of Sir John Cockcroft and were known as 'Cockcroft's Folly'. As it turned out, this afterthought averted a catastrophe because the filters trapped a large proportion of the particulate material from the fire and only failed to stop the emission of gases and fine particulates. These included the fission product iodine-131 and, as it turned out later, polonium-210, which is also highly volatile.

Even though the extent of the fire and the seriousness of the situation was realized, the public was not informed. The Windscale staff had no real idea how to quell the fire now that normal air cooling was inappropriate; during the fire, 11 tonnes of uranium were ablaze. Carbon dioxide was tried, but the temperature of the fire was so high that this merely fanned the blaze. As a last resort it was decided to try water although this was not an ideal choice because of the possibility of reactions that might produce explosive hydrogen. This had happened at a reactor accident at Chalk River in Canada in 1952 and was to reoccur, without explosion, at Three Mile Island in the USA in 1979, and, with a devastating explosion, at Chernobyl in 1986. But something had to be done to quell the Windscale fire, so the Chief Constable of Cumberland was alerted to the possibility of an emergency (although presumably one had already occurred) and at 8.55 am on Friday 11 October, fire hoses were coupled to the reactor and the core was flooded. This approach was successful and the fire subsided, but this was nowhere near the end of the problem. A vast cloud of radioactive fission products had been released, dominated by 20,000 curies (740 TBq) of iodine-131. Decisions had to be made about information that was to be given to the public about their exposure to radiation. This must have been complicated by the secrecy surrounding Windscale at the time. Windscale No. 1 and No. 2 reactors were nominally designed to produce plutonium for the weapons programme but they were also used for neutron irradiation purposes. At the time of the fire there were several sealed capsules within the reactor. The release of these materials, notably polonium-210 which was used as a trigger in nuclear weapons, plutonium-239 and tritium, was not recorded in official reports at the time and its impact was only really assessed in 1983 by the NRPB. Interestingly, although there were measurements of polonium-210 and plutonium-239 on filter papers and in biota in the UK, the only measurement of tritium was from Hamburg in West Germany on the 11 and 12 October. Total releases deduced by the NRPB in their reassessment are given in Table 4.1. This reassessment, in the light of full knowledge, increased the calculated collective effective dose equivalent commitment of the UK population by 67 per cent to 2000 man-sieverts.

The government's reaction to the incident (apart from an attempted hush-up) was to introduce a ban on milk consumption at a level of iodine-131 of 3700 Bq per litre. This ban covered Cumberland and Westmorland, where the highest deposition of

Table 4.1 Estimates of Amounts of Radionuclides Released in the 1957 Windscale Fire

Nuclide	*Quantity released (TBq)*
I-131	$7.4 \times 10^2 - 1.2 \times 10^3$
Po-210	8.8
Pu-239	1.6×10^{-3}
U-238	6.0×10^{-5}
U-235	2.0×10^{-6}
H-3	5.0×10^3
Xe-133	1.0×10^4

iodine-131 had been 960 kBq/m^2 (compared with 15 kBq/m^2 in the same part of the country after Chernobyl), and other parts of northwest England, an area of 500 sq km. The ban had the effect of reducing the dose to children's thyroids (where iodine concentrates) by a factor of five. The milk withheld from sale was poured away into rivers and streams without any consideration of the possibility of producing dried milk from it after iodine-131 decay. This ban on milk consumption seemed to convince the public that the incident was under control, a feeling fed by the conclusions of the official report of the incident (Cmnd 302) published in November 1957. The report only mentioned iodine-131, strontium-89/90 and caesium-137 as having been found in the environment and, after having estimated maximum likely child thyroid doses of 25 rad (0.25 Sv) in the control area, concluded that 'it is in the highest degree unlikely that any harm has been done to the health of anybody'. This statement is in stark contrast to the conclusions of the two NRPB reports published in 1982–3 after all the facts of the environmental impact of the fire had been finally released. The NRPB calculated that there might be about 250 cancers and about 30 extra cancer deaths in the years from 1970 onwards. This probable mortality figure was of course calculated using the then accepted ICRP risk factors. Clearly, the contamination from the fire had been much more widespread than the government was prepared to admit: 200-million people would have had detectable intakes of iodine-131 delivering a total collective dose of 20,000 man-sieverts.

We seem to have learnt little from the incident as far as emergency planning is concerned, as was seen in the chaos following Chernobyl in 1986. The two Windscale reactors were entombed and remain so to this day. However, it could be argued that one good thing to come out of all of this was the creation of the Nuclear Installations Inspectorate (NII) in 1959.

Three Mile Island

Brief details of the accident at Three Mile Island are included here because it was a turning point in the history of the use of nuclear power. There seems to be considerable uncertainty over the *actual* releases of radioactive materials to the environment and the concomitant harm that may be experienced by the local population. However, there is no uncertainty that there was the potential for the accident to have developed into a catastrophe.

Three Mile Island II is one of two pressurized water reactors (PWR) on an island in the Shenandoah River near Harrisburg, Pennsylvania. On 28 March 1979 the water supply to one of the steam generators failed. The turbines (turbo-alternators) cut out and the reactor was shut down by automatic insertion of the control rods. An alternative boiler feed pump should have started up but failed to do so because valves that should have been open had been left closed after routine testing. The reactor was now in an unstable state because it was not being cooled and so started to heat up. A second valve failure allowed steam to escape and the emergency core-cooling system automatically came into operation. Although some valves had failed, up to this point the automatic protection systems of the reactor had worked reasonably well. However, at this point there was human intervention. An operator, believing the pressure vessel to be full of water, turned off the core-cooling system, thus allowing some of the fuel to melt. This fuel reacted with steam and water and produced a bubble of hydrogen gas in the primary containment. There was also a release of radioactive fission product gases (krypton, xenon and about 600 MBq of iodine-131) to the environment. Although alarms had begun to flash in the control room in increasing numbers, the operators were still not sure what was wrong. The main coolant pumps were now pumping a mixture of steam and water and shaking violently. It was only after frantic consultations among senior staff of Metropolitan Edison (the reactor operators) and Babcock & Wilson (the reactor builders) that the situation was appreciated and action taken (closure of the pressure relief valve, for instance) to stabilize the reactor. Nevertheless a tense three days followed during which the potentially explosive hydrogen bubble above the reactor core occupied everyones' minds. Eventually the bubble dispersed of its own accord but not before there had been voluntary evacuation of some of the surrounding populations. This had been encouraged by the Governor of Pennsylvania when he failed to get consistent or reliable advice from reactor 'experts' on the potential for disaster.

The accident probably resulted in very little release of radioactive materials off-site and therefore insignificant long-term health effects. However, it shocked US public opinion and was one reason for the slowing down, if not the demise, of the US nuclear industry. Control room chaos and poorly understood instrumentation were officially blamed for the incident but the confusion and contradictory advice that followed cast the industry in a very poor light. Two reports, the Kemeny Commission report and the Rogovin report of the Nuclear Regulatory Commission of 1980, painted a very disturbing picture of total chaos and potential disaster. Although the reactor owners protested that the automatic safety features had saved the day, it is noteworthy that no new reactors have been ordered in the USA since 1978.

It was noticeable that in the case of the Three Mile Island accident the pro-nuclear lobby emphasized the fact that there had been no significant releases of radioactive materials, whereas anti-nuclear pressure groups pointed to the fact that potential disaster had apparently been averted by luck. For some reason, which is incomprehensible, this sort of accident was considered of low probability and no emergency procedures had been rehearsed or were able to cope. Unfortunately, the lesson was learnt too late and Three Mile Island had a devastating effect on the US nuclear industry.

Information and Credibility

Without information, it certainly is not apparent to the general public that they are being irradiated or that their environment is being polluted. Radiation impinges on none of our five senses and is therefore understandably regarded as more insidious than other environmental pollutants, some of which can at least be seen or smelt. Reassurance that the environment is unpolluted and that food and water are 'safe' to consume depends on the credibility of the individual or organization doing the reassuring. The government lacks that credibility and is never likely to acquire it if its recent efforts are taken as typical. This lack of confidence in government edicts on radiation matters has coincided, not surprisingly, with an increasing awareness of environmental issues. Part of this awareness has been heightened by the growth in stature of international pressure groups (led by Friends of the Earth and Greenpeace) which, as they have acquired more financial support have increased their integrity and responsibility. Because of their investigations incidents of pollution have been revealed that might have remained undetected until the cause of a long-term health effect was being sought in say thirty to forty years' time. However, it is a pity the public has to rely on pressure groups because the credibility of the government is so low. After all, government agencies have, or should have the resources to provide blanket monitoring and surveillance facilities.

One of the worst examples of governmental chaos, resulting in a massive credibility gap, was the reaction to the Chernobyl disaster in 1986. In this particular case official inconsistency was not restricted to the UK, but the fact that no two countries in Europe could agree on a radiation surveillance policy or intervention levels of contamination was, to say the least, worrying. The aftermath of the Chernobyl incident will therefore be explored in some detail here, particularly with regard to official pronouncements over food contamination, to illustrate the chaos at the time and anxieties for the future.

The Chernobyl Incident and After

From the point of view of public confidence, it is worth exploring what exactly went wrong at the Chernobyl nuclear power station in 1986 and attempting to learn from the mistakes made. Probably the most important point to arise from this disaster is that, bearing in mind that radioactive fallout has no respect for national frontiers, continuous assertions that 'it couldn't happen here' are only of marginal value.

Chernobyl is a small town in the eastern part of Byelorussia-Ukraine about 600 km southwest of Moscow (see Figure 4.1). It had a population of 12,500 in 1986 and is near the confluence of the Pripyat and Dnieper rivers. These feed a large lake/reservoir about 110 km long which supplies water for the region, in particular Kiev, which is 130 km south of Chernobyl. The possible contamination of this reservoir was a continuing anxiety in the weeks and months after the incident.

The Chernobyl nuclear power station is about 12 km up river near Pripyat, a small town (population 45,000) built essentially to house workers from the power station site. The power station comprised six RBMK-1000 reactors, four operational and two under construction. The RBMK reactor is unique to the Soviet

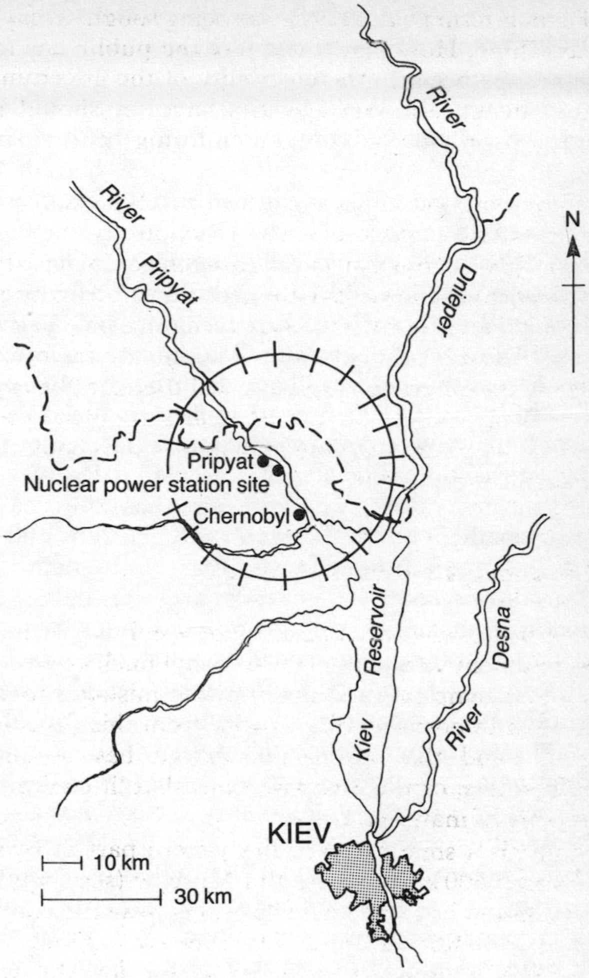

Figure 4.1 Location of the Chernobyl Nuclear Power Station.

Union, being a hybrid of three other designs. It uses graphite blocks as a moderator and water that is allowed to boil in pressure tubes as a coolant. This type of 1000 MW reactor is quite common in the USSR, there are at least 28 either operating or under construction. The design has also been stretched to an output of 1500 MW at stations at Ignolina in Lithuania. Since the accident the RBMK design has been subjected to critical scrutiny by the nuclear industry in the West. Bearing in mind the international nature of the fallout contamination from the incident, it is worrying to hear that the design, although heralded as the pinnacle of safety and efficiency in the Soviet Union, would not have been licensed to operate in the UK.

The RBMK reactor is physically impressive. The 2000-tonne graphite core is 7 m high and 12 m in diameter. The fuel is about 180 tonnes of 2 per cent enriched uranium (i.e. uranium-238 enriched with 2 per cent of uranium-235) fabricated into rods and sleeved in zirconium cladding. These fuel rods or pins are loaded coaxially into nearly 1700 pressure tubes in the core through which a mixture of water and steam circulates to remove the heat of the fission reaction. The steam is used to drive two 500 MW turbo-electric generators. The core also has 200 channels for control rods made of boron, which absorb neutrons and are used to control the uranium fission chain reaction by being raised or lowered into the core. These rods are normally positioned mechanically, but under emergency conditions they may be dropped under gravity. The control of the rods was to have a crucial bearing on the events that led up to the accident.

The graphite core is encased in a steel-lined reinforced concrete building filled with inert helium-nitrogen which is important in preventing spontaneous combustion of the graphite (at 700°C under normal operation). This primary encasement also reduces radiation doses to personnel on the outside. However, no secondary containment was provided and although this inadequacy was criticized afterwards, it is significant that some British reactors have no secondary containment either. A form of secondary concrete outer 'dome' encasing the whole reactor might have prevented some of the release of radioactive material.

One technical feature of this sort of reactor is the use of some of the power generated to drive the reactor coolant pumps and other safety systems. This is backed up by a standby diesel-electric power supply that switches on automatically should power fail. However, should steam pressure to the main turbogenerator fail the turbine will continue to spin freely for several minutes. During this freewheeling period power should still be generated to power coolant pumps and other vital plant. Technicians were scheduled to investigate this effect after creating the conditions for a steam pressure failure at the end of April 1986.

At 1.00 am on Friday, 25 April 1986, technicians started the annual maintenance and tests of the reactor including a free-wheeling rotor test. The power on No. 4 reactor was cut by half and one of the two turbogenerators was switched off, its steam supply and circulation pumps being switched over to the other. After 13 hours, the emergency core-cooling system (ECCS) was switched off in order to prevent it automatically tripping-in during the subsequent tests. At this point the Kiev controller asked for more power to be supplied to the grid and all tests were postponed. However, the ECCS was not reconnected and for the next nine hours the reactor produced power without this automatic protection system. This alone was a flagrant disregard of safety rules that should not have been physically possible.

Later that evening (at 23.10 pm on 25 April), the power generation requirement was lifted and tests were resumed. In a fresh attempt to lower the power output the technicians switched to manual control. Unfortunately they failed to control the drop of power and it dropped to about 30 MW, well below the minimum level for control. This was a serious situation in this design of reactor and the technicians realized it. After two hours of feverish work (at 1.00 am on 26 April), they managed to lift the power output to 200 MW. A major difficulty was the build-up of a fission product gas, xenon. This is a neutron absorber and therefore poisons the fission reaction. In order to overcome the xenon effect the technicians had withdrawn a large number of control rods and the reactor was now very unstable. Unbelievably, at 1.17 am, the experiment was then continued. Six of the main coolant pumps were switched on, followed immediately by another two. These produced an enormous flow of water through the reactor that reduced the reactivity even further, since water is a neutron absorber. In an attempt to counteract this seesawing of power, more control rods were removed and further emergency control rod reinsertion systems overidden. At 1.23 am, steam was present in some parts of the reactor and water in others. With bubbles of steam forming, less cooling and neutron absorption was occurring and the power was about to surge. Nevertheless the experiment proceeded; the turbine warning systems were turned off and finally the steam valve to the turbine was shut. The test conditions had now been created and the turbine freewheeled to rest. This meant that less and less power was supplied to the coolant pumps, causing a greatly diminished flow of coolant, increased steam pressure and a rise in reactivity and heat. The 'kettle was boiling dry' and without control facilities was about to self-destruct.

At this eleventh hour the technicians frantically tried to regain control, but only six of the minimum thirty required control rods remained in their positions. It was too late; the core channels were distorting due to overheating and the rods would not even drop under gravity into the core. At 1.24 am the core temperature rose to 3500°C (five times the normal operating temperature) and an explosion was heard by the control room staff. This was caused by coolant reacting with the red-hot zirconium cladding to produce hydrogen. This in turn produced a large overpressure explosion through the reactor building's airtight seal and allowed the entry of air, causing a second, much greater explosion. It appears that the escaping steam contacted red-hot graphite, producing an explosive mixture of water gas (carbon monoxide and hydrogen). The second explosion blew apart the core, destroyed the reactor building and lifted aside the 1000-tonne steel floor plate. The huge fuel-loading machine crashed into the now burning reactor core. A column of fire, smoke, hot graphite and radioactive material erupted into the night to a height of 1200 m and the contamination of the whole of Europe had begun.

There followed a chain of events that resulted in the deaths, in horrible circumstances, of thirty-one people within a few weeks, the hospitalization of more than 200, who recovered only slowly, and the evacuation of 135,000. The reactor fire burned for ten days before it was quenched, threatening at some stages to melt through its base (the so-called 'China Syndrome'). During this time it poured an incredible amount of intensely radioactive material (2,000,000,000,000,000,000 Bq or 2×10^{18} Bq) into the atmosphere. About half of this material was strewn across the Ukraine and the rest drifted towards Western Europe.

The contamination of Europe

The first evidence of the arrival of the radioactive cloud outside the Soviet Union was contamination on the clothes of workers passing monitors at the Forsmark nuclear power station in Sweden at 9.30 am on Monday, 28 April. At first, the Swedes thought the contamination had come from their own reactor and a quick survey of the site indeed revealed generalized doses up to about 100 times normal background. The fact that the source of the contamination was outside Sweden was soon revealed by a study of the records of remote monitoring stations. Unlike the UK, Sweden has a nationwide network of gamma monitors which, ironically, are not manned at weekends. However, the automatically recorded counts showed that the radioactive material had first arrived at 14.00 on Sunday, 27 April. These results were soon confirmed by reports from Finland, Norway and Denmark. At 19.00 on 28 April, some sixty-six hours after the incident had occurred, the USSR admitted that No. 4 reactor at Chernobyl had been destroyed by an explosion and was on fire.

It took another four days for the cloud, now considerably depleted after a tortuous path across Europe (see Figures 4.2 and 4.3), to arrive over southeast England (Friday, 2 May). It then proceeded north and west to meet that weekend's rain. 'Rain-out' of particles from the cloud took place over the next two days, principally in areas that received the heaviest rain such as North Wales, Cumbria and southwest Scotland, but other areas also received some contamination (Figure 4.4).

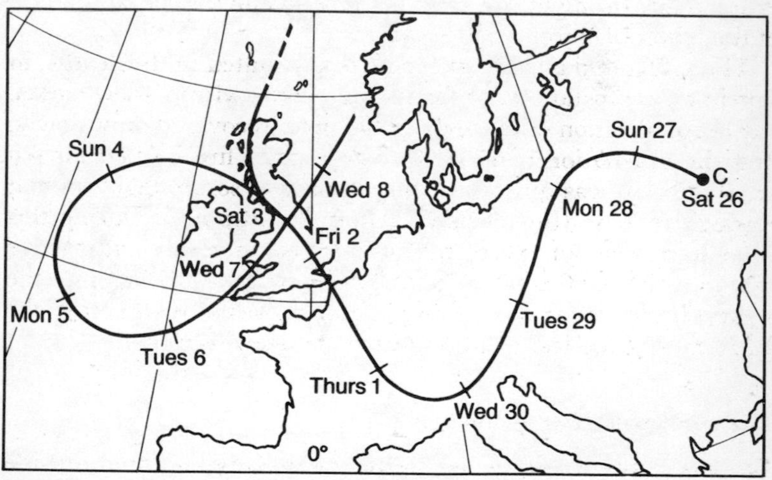

Figure 4.2 Estimated Path of the Chernobyl Radioactive Cloud that Ultimately Crossed the British Isles (26 April–8 May 1986).

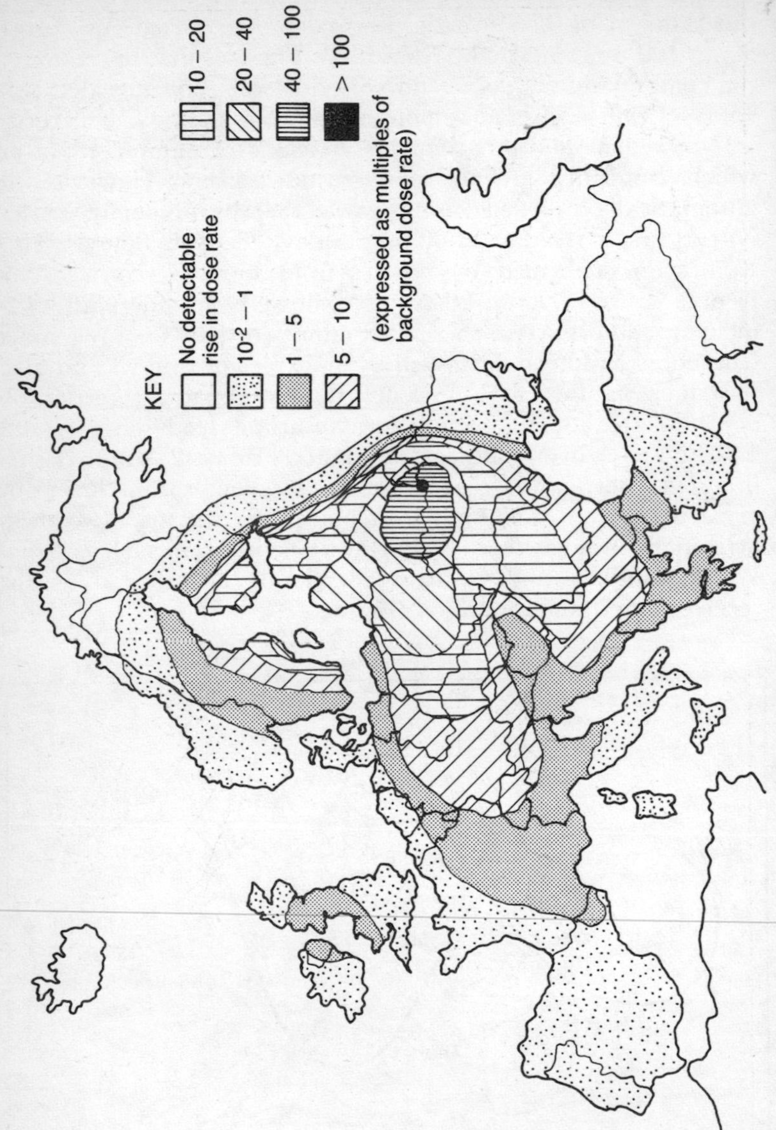

Figure 4.3 Gamma Background Across Europe on 3 May 1986.

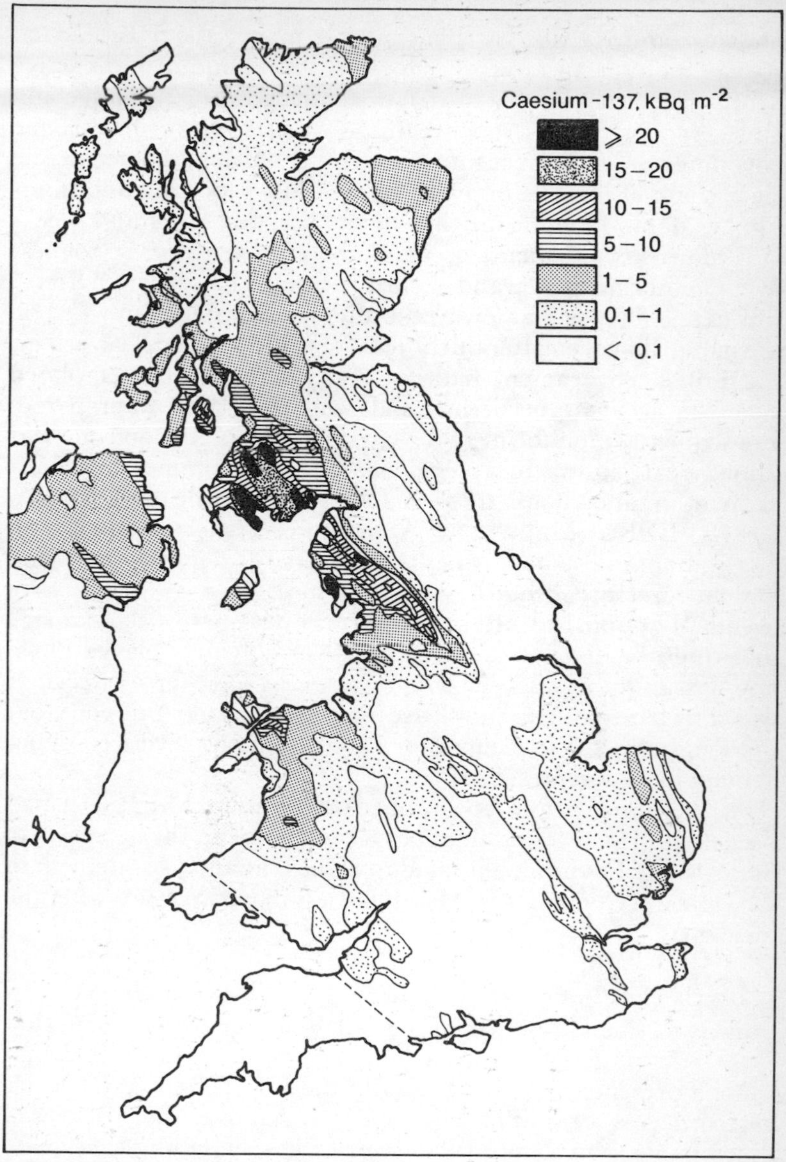

Caesium-137, kBq m^{-2}

$\geqslant$ 20

15–20

10–15

5–10

1–5

0.1–1

< 0.1

Figure 4.4 Estimated Deposition of Caesium-137 in the UK from Chernobyl (in 10^3 Bq/m^2).

UK government reaction

Unlike countries like Sweden and Germany, the UK had no remote monitoring network. However, the cloud of radioactive debris took nearly a week to arrive over the UK and its path could have been predicted from weather forecasts. Nevertheless, the government was thrown into a state of confusion. There was no clear chain of responsibility for monitoring, reporting and interpreting data. This meant that in the first few days and weeks government departments oscillated between frustrating coyness over monitoring results and absurd releases of handwritten lists of raw data. In those first few weeks after the disaster, it dawned on the public that even after thirty years of worldwide nuclear power, the British government had not really seriously contemplated such an accident occurring and was singularly unprepared. Consequently monitoring of the environment, biota, food etc. was initially carried out in a very uncoordinated manner by a vast array of organizations; these included the NRPB, the UKAEA, CEGB, BNFL, Ministry of Agriculture, Fisheries and Food, hospital physics departments, university departments and various pressure groups. Although most of these data were eventually published, the government, ever wary of the damage that could be done to the nuclear power industry, played down any possibility of lasting contamination or long-term effects. It was thus left to others (not least the present author) to point out the long-term consequences for health that the levels of radioactivity in food represented.

At this time, before any proper or considered appraisal had been done of food contamination, Kenneth Baker, the Secretary of State for the Environment, made a by now infamous statement in the House of Commons (6 May 1986, less than two weeks after the incident).

> Levels of radiation are now likely to decline over the next few days . . . The House will appreciate from what I have said that the effects of the cloud have already been assessed and that none presents a risk to health in the UK. The highest levels recorded as a result of the cloud occurred over the weekend of 3rd to 5th May. Since then levels have been falling every day, and are now either at or approaching normal background levels in all parts of the country. As long as there are no further

discharges from Chernobyl, the incident may be regarded as over for this country by the end of the week although its traces will remain.

As an example of a totally misleading statement this takes some beating. It could be argued that the government was receiving poor advice but there is evidence, freely available in the scientific press, that indicates that certain fission products (Cs-137) are retained in organic, peaty soils for years. These data, which were accrued from experiences with fallout from weapons testing, were published in the early 1960s by the Agriculture Research Council. Had this evidence been taken into account, it is to be hoped that such bland reassurances would not have been given. This is especially true with hindsight because it was already apparent to the MAFF that the problem of caesium was not going to go away. The levels in sheep meat from Cumbria and North Wales were well above the intervention level of 1000 Bq/kg recommended by the 'Article 31 group of experts in Europe' (a group of scientists called together to advise the EEC under Article 31 of the Euratom Treaty). This sort of information had already been issued to the public in the Department of the Environment (DOE) first compilation of monitoring data (July 1986). In fact, the situation was becoming even more confused because, although higher than acceptable levels of caesium-137 and caesium-134 were known to be present in lamb, nothing was done. Michael Jopling, the Minister of Agriculture, made the following statement to the House on 30 May:

> The results of the monitoring show that levels of caesium-137 and 134 in these products [milk products] are generally low but higher readings have been found in the meat of some hill sheep and lambs (which are not ready for slaughter at this time of year) in upland areas of the country which were subject to high rainfall over the weekend 2–3 May. These levels do not warrant any specific action at present.

Subsequently, this statement was found to be less than accurate because some lambs had already gone to market (over 37,000 according to Labour Party researchers). Eventually the government was forced to do something and, about five weeks after Baker's assurances that everything would shortly go away, a restriction was placed on the slaughter and sale of lamb containing more than 1000 Bq/kg of caesium 137/134. This was accompanied by the following statement by Jopling in the House on 20 June 1986:

> The monitoring of young unfinished lambs not yet ready for market in certain areas of Cumbria and North Wales indicates higher levels of radio-caesium than in the rest of the country. These are the areas of high rainfall during the weekend of 2/3 May. While these levels will diminish before the animals are marketed, My Rt Honourable Friend the Secretary of State for Wales and I have decided to use the powers in the Food and Environment Protection Act 1985 to make absolutely certain that when these lambs are marketed, they will be below the internationally recommended levels for radio-caesium of 1000 Bq/kg.

There has followed a restriction that has now (in 1989) lasted three years.

Intervention levels

It is pertinent at this point to look more closely at the derivation of the 'internationally recommended level' that Michael Jopling mentioned in his statement and to judge how much confidence is engendered by the way these levels are derived.

Levels of 'intervention' or 'emergency reference levels' are derived levels at which some action is thought to be required immediately after an accident or incident. Because the increased contamination is not expected to last for very long, the derived limits are either for surface contamination or for foodstuffs for which the incorporation time is short, e.g. milk, drinking water and green vegetables. The actions or interventions anticipated are designed to limit the dose and therefore the damage to health of the most susceptible groups in the population. These actions might include evacuation, sheltering, issue of iodate tablets for thyroid blocking (i.e. prevention of iodine uptake), or banning of certain foods. The last of these options was pertinent in the UK after the Chernobyl incident, and was considered by the government. In the event, no action was taken relative to the derived emergency reference levels (DERLs). This was almost certainly because of issues other than pure science. For instance, DERLs are set at two levels, an upper level that corresponds to a whole body dose to the individual of 50 mSv, and a lower level of 5 mSv. At levels of contamination that could correspond to doses above the upper level, action must be taken. Above the lower limit but beneath the upper level, action should be considered, but only taken if food supplies and/or nutritional needs would be unaffected. Scenarios that could involve panic might also limit options in this region, but certainly value judgements have to be made. Below the lower level, actions or intervention is not considered necessary. These recommendations were made by the ICRP in its Publication 40 and endorsed by the NRPB most recently in its report R-182, just prior to Chernobyl in March 1986. In this report the NRPB gave DERL values for a number of radioisotopes. Values are given in Table 4.2 for I-131 and Cs-137.

Clearly, models involving assumptions about food intake, metabolism etc. must be used to derive these figures and generally they seem conservative, although the time over which they could be in force is not so clear. The risk at the upper level is that associated with about ten times the legal annual public dose limit (5 mSv) in force at the time. Similar models are utilized in deriving another set of guidance levels known as Generalized Derived Limits (GDLs). These serve an entirely different purpose. They were defined in ICRP Publication 26 which stated that 'the intention should be to establish a figure such that adherence to it will provide virtual certainty of compliance with the Commission's recommended dose equivalent limits'. Thus GDLs are linked directly to dose limits and for contamination of

Table 4.2 DERL Values for Iodine-131 and Caesium-137 Given by the NRPB

Milk *Bq/l*	*Drinking water* *Bq/l*	*Green vegetables* *Bq/kg*
Iodine-131		
Upper 20,000	24,000	1,100,000
Lower 2,000	2,400	110,000
Caesium-137		
Upper 36,000	73,000	1,900,000
Lower 3,600	7,300	190,000

the environment, food chains etc. are to be considered as a means of checking compliance with the public dose limit. These limits are used for continued contamination of the environment as could derive from, say, discharges from a nuclear site or the long-term aftermath of an accident such as Chernobyl. Again 'critical' groups of the population are considered: that is groups who may be more susceptible or, because of their dietary habits, more at risk. The groups generally considered separately are adults, 10-year-old children, and year-old infants. The dose is integrated over fifty years to give the committed dose, and the most limiting GDL is quoted, e.g. if the GDL for 10-year-old children is limiting, the other two groups will be overprotected. Unfortunately, the recommendations over the public dose limit were changed in 1987 and thus there are at least two sets of GDLs to consider in relation to the aftermath of Chernobyl (1986). Some selected values recommended by the NRPB for caesium are given in Table 4.3.

There must, of course, be many uncertainties in the derivation of the figures quoted in Table 4.3 and many points over which the assumptions might be attacked. Nevertheless they represent a

Table 4.3 NRPB Recommended GDL Values for Caesium (Bq/kg unless otherwise stated)

	1983–87	*After Aug. 1987*
Caesium-134		
Milk (Bq/litre)	2,000	200*
Milk products	4,000	1,000*
Beef	4,000	1,000†
Mutton and lamb	8,000	2,000†
Green vegetables	2,000	700*
Drinking water (Bq/m^3)	200,000	100,000
Caesium-137		
Milk (Bq/litre)	3,000	300*
Milk products	4,000	2,000*
Beef	5,000	1,000†
Mutton and lamb	10,000	3,000†
Green vegetables	2,000	1,000*
Drinking water (Bq/m^3)	300,000	100,000

Note Critical group ages: *1-year-old infant; †10-year-old children

reasonably scientific approach to the estimation of risk with a published set of assumptions. However, the abrupt change in 1987 and the apparent disagreement with the 'internationally agreed limit' quoted by the minister should give some cause for concern.

The caesium that arrived in the UK from Chernobyl had a very characteristic 'fingerprint'. The ratio of activities of caesium-137 (a fission product) to caesium-134 (produced by neutron activation of caesium-133) was 2:1. Subsequently, this ratio altered with time, as caesium-134 decays with a half time of 2.5 years. Currently (1989), the ratio is nearer 5:1. This distinguishes Chernobyl caesium from, say, that discharged from Sellafield, for which the ratio is 20:1 or higher. Thus the current GDL for Chernobyl caesium in lamb should be 2830 Bq/kg and would have been about 9300 Bq/kg in 1986.

Why then does not the limiting figure of 1000 Bq/kg (for combined activities of both isotopes of caesium) quoted by the Minister of Agriculture in June 1986 for lamb (which still applies today) appear in these tables? The answer to this question has partly to do with conservatism but largely to do with politics rather than science. It was thought that people might eat more than one foodstuff contaminated with caesium and, as GDLs strictly apply to one food, their application was thought to be inappropriate. This situation was certainly true soon after Chernobyl but at longer times nearly all radioactive caesium entered the food chain in meat. It is also difficult to envisage how GDLs might ever be applied in practice. In addition, the quantum leap downwards from over 9000 Bq/kg (see Table 4.3 and calculation above) to 1000 Bq/kg needs some explanation. A more complete answer comes from considerations of the impact of Chernobyl on EEC trade policies.

The reaction of other European countries to Chernobyl

The reaction of Western European countries, particularly those in the EEC, to the fallout from Chernobyl shows confusion and contradictory measures. There were common features but also a great deal of diversity that cannot be explained solely in terms of radiation levels. Southern Germany, for example, was one of the worst-affected EEC areas. In most of West Germany action was taken, but not uniformly. The Federal government set a maximum level for iodine in milk of 250 Bq/litre but most of the Lander (states) set their own, e.g. Hessen and Hamburg's were as low as 20 Bq/litre. Initially, the Federal government had advised people not to drink fresh milk, but this was retracted on 6 May. On the same day the state of Nordrhein-Westfalen halted all school milk supplies.

Green vegetables were also controlled variously, i.e. they were either confiscated, ordered not to be sold or listed as items that should not be consumed. Milk was available in Bavaria with an iodine level of 900 Bq/litre but the authorities in Berlin destroyed 24,000 litres of East German milk that was eventually found to contain a concentration of 7.5 Bq/litre. Vehicles and food were checked for contamination if they came from the East but it was ten days before it was realized that contamination could also be imported from such places in the south as Italy.

In France, there was a situation of silence from the authorities and therefore ignorance for the population. It was eventually announced that while all neighbouring countries had been substantially contaminated, the slight increase in background in France would have no impact. Later, spinach was banned from Alsace and a unilateral ban was placed on food imports from Eastern Europe.

In other countries less affected by fallout, actions taken by governments were again inconsistent and generally the information that was given out was obscure. This was particularly apparent in Belgium and Greece. Other examples of 'action levels' are listed in Table 4.4.

These variations in reaction to this incident are extremely disquieting. There is obviously more than just science involved in the setting of limits but it must have been bewildering for inhabitants of frontier districts, who experienced some of the more extreme contradictions. People in Alsace could observe total bans on the consumption of certain foodstuffs and mass crop destruction on the German side of the Rhine while a perfectly normal way of life went on a kilometre away on the French side. A similar situation occurred around Lake Constance; the Germans destroyed their crops and the Swiss did not. Food in Italian Sardinia was regarded as contaminated but this was apparently not the case in nearby French Corsica. It looked as if there was a national variation in sensitivity towards radiation! However, it was significant that those countries with the greatest dependency on

Table 4.4 Examples of Post-Chernobyl 'Action Levels' for Caesium

Country	*Food*	*Action level (Bq/l or Bq/kg)*
Brazil (Cs-137 + Cs-134)	Milk powder	3,700
	Other foods	600
Canada (Cs-137)	Milk	50
	Dairy products	100
	Other foods	300
	Spices	3,000
EEC (Cs-137 + Cs-134)	Milk	370
	Other foods	600
China (Cs-137)	Milk	4,600
	Fruit and vegetables	1,000
	Cereals	1,200
	Beverages	460
Sweden (Cs-137)	All foods	300
USA (Cs-137 + Cs-134)	All foods	370

nuclear power tended to do the least, e.g. France and Belgium, and people began to wonder what role their personal safety played in the matter.

Given these differences in approach, it was some time before the EEC could react effectively. It was quick to produce proposals on intervention and banning levels in food but these were just as quickly rejected by nations attempting to protect their own national trade interests. The Italians, for example, wished to protect their vegetable trade, and the Germans their milk exports. Very roughly the recommendations of the Community came in three stages and as the UK, at least, attempted to follow them, it is worth considering them.

Stage one was the issue of a recommendation on 6 May 1986 on the maximum permissible concentration of radionuclides in milk, fresh fruit and vegetables. The second stage was a ban on imports of fresh food from Eastern Europe and the third was the recommendation of maximum levels in foodstuffs from all third countries.

The history of these recommendations and decisions made by the EEC after Chernobyl makes for interesting study. The first recommendations (86/156/EEC) referred to maximum activities in respect of marketing and export within the Community. These maxima for iodine-131 between 6 and 26 May are shown in

Table 4.5 (these change as the iodine-131 decays with a half life of eight days). The Commission also banned certain imports from Eastern Europe, including fresh meat until 31 May.

On 12 May, regulation 1388/86, which suspended imports of agricultural products originating in Bulgaria, Czechoslovakia, Hungary, Poland, Romania, Russia and Yugoslavia, came into force. This was subsequently superseded by regulation 1707/86 of 30 May in which maximum permissible radioactivities were stipulated for import of certain foodstuffs from all third countries. For caesium-137 and caesium-134, these were 370 Bq/L or L/kg for milk and baby food and 600 Bq/kg for all other foodstuffs. These levels were specifically to apply to inter-Community trade. However, at that time the UK was operating an intervention 'banning level' of 1000 Bq/kg for caesium-137 plus caesium-134 in foodstuffs, particularly lamb. This level had been set by the 'Article 31 Expert Group' of the EEC at a meeting in Luxembourg on 22–23 May 1986. The exact scientific rationale behind the derivation of this level is of great significance as it has now been used for banning sales of lamb for more than three years in the UK.

The action level for caesium

The 'internationally recommended action level for caesium' in food, as Michael Jopling called it on 20 June 1986, has some rather obscure antecedents. It was international in that it was agreed by the EEC Article 31 group of experts but several countries were already applying their own intervention levels. Sweden, for instance, was using a level of 300 Bq/kg three weeks after the incident. However, it was intended to be used in the EEC countries for major foodstuffs, and the UK adopted it immedi-

Table 4.5 Maximum Activity of Iodine-131 (Bq/kg) Laid Down by EEC, 1986

	Milk and milk products	*Fruit and vegetables*
6 May	500	350
16 May	250	175
26 May	125	90

ately. The starting point of the calculations are the annual limits of intake (for ALIs see Chapter 3) for caesium isotopes recommended by the ICRP in Publication 30 (1980).

The values (for annual intake by adults) were 300,000 Bq for caesium-134 and 400,000 Bq for caesium-137. These limits were derived from models based on a maximum annual dose for members of the public of 5 mSv. By 1986 the ICRP and the NRPB considered that the average annual dose should not exceed 1 mSv over a lifetime, and as this food contamination was expected to persist, these intakes were divided by five. In addition, it was recognized that the critical group, which would receive the most significant dose, consisted of 10-year-old children. Because caesium is distributed fairly uniformly in the body and, except for very young children, its retention time is independent of age, the intakes may be scaled by body weight, i.e. divided by a factor of two. Thus, dividing by these two factors, maximum annual intakes of caesium for 10-year-old children were calculated to be 30,000 Bq for caesium-134 and 40,000 Bq for caesium-137. It should be noted that this already contradicts the NRPB's report DL7 which was published three years previously. In this report on generalized derived limits (GDLs) the annual limits of intake for ingestion of caesium isotopes for 10-year-old children were, when scaled for an annual dose of 1 mSv, six times higher (less restrictive) for caesium-134 and five times higher for caesium-137.

These intakes by ingestion were then converted into concentrations in food. It was considered that children eat more food in relation to their body weight than adults. The range of food intake of a European 10-year-old was taken as 1–1.5 kg/day, an unknown fraction of which might be from the contaminated zone. It was thought that control or intervention levels must be based on daily intake of contaminated food over long periods. Five per cent of the diet of such contaminated food (25 kg/year) was considered an appropriate fraction. Thus, these factors give intervention levels of 1200 Bq/kg for caesium-134 and 1600 Bq/kg for caesium-137. The precision of these values was considered insufficient to justify the use of two significant figures and a single value of the mixture of the two nuclides of 1000 Bq/kg was to be used. This figure was far more restrictive than the figures of 1600 Bq/kg for caesium-134 and 2000 Bq/kg for caesium-137 for lamb intake (when scaled for a dose of 1 mSv per year) as recommended by the NRPB (see Table 4.3).

It can be seen that this level is fairly conservative when set against the risk rates in use at the time. Part of this conservatism stems from the dietary fraction derived from the contaminated area. In later predictions of whole-body retention of caesium in adults, estimates derived from the assumptions that 100 per cent of diet comes from local produce (in Cumbria) were shown to be about five to six times too high when the individuals were counted in a whole-body counter. Clearly this is an important factor and, if contamination were more widespread in a future discharge, would require reappraisal.

This restriction level could also be considered conservative because the *actual* dose commitment received by, say, a 10-year-old child eating 20 kg of lamb per year, would be small (about 0.28 mSv). This would give a yearly risk of cancer of just over 1 in 350,000 (using current ICRP risk rates) or about 1 in 80,000 using more contemporary cancer risk rates.

International food contamination limits

It is certainly clear that the UK government, in common with other governments in Europe, was faced with a dilemma because its internal control levels of radioactivity in food were vastly different from those of neighbouring countries and those recommended by the EEC. The reasons for these differences lie in varying attempts to protect trade rather than protection of health. Nevertheless, the confusion this caused, and still causes, has increased suspicions that such levels are *not* scientifically based and that some populations are being protected more than others. Certainly the inconsistencies even within the UK when the government chose to ignore the NRPB's recommended limits on 22 May should be explained. Perhaps the most significant of the EEC pontifications occurred as a result of another meeting of the EEC experts in Luxembourg between 27 and 30 April 1987 to agree intervention levels to be used in future emergencies. The levels agreed are given in Table 4.6.

At the time of this meeting, the UK still had a meat banning level of 1000 Bq/kg in operation. As can be seen from Table 4.6, the 'scientific experts' recommended a level of 5000 Bq/kg. Clearly something had to be done quickly to bridge this gap between 'science' and government credibility. Thus other factors – 'public and political acceptability' – were deemed to be of

Table 4.6 EEC Intervention Levels Agreed in April 1987 (all figures in Bq/kg)

Radionuclides	*Milk/infant food*	*Other major foods*	*Water*
Iodine/Strontium	500	3,000	400
Plutonium (etc.)	20	80	10
Caesium (etc.)	4,000	5,000	800

overriding importance and the recommendations of the 'experts' were divided by various factors to give the values for imports, home production and intra-EC trade listed in Table 4.7. It is notable from Table 4.7 that natural and man-made radionuclides are not distinguished, e.g. is naturally occurring K-40 to be included in the fourth category with Cs-134/Cs-137?

It is clear from all of this that there is no such thing as a 'scientifically' based dose or contamination limit for food. Any limit is a trade-off between future health detriment and cost, and immediate social, political and economic disruption. On 28 October 1987, the UK government passed a resolution urging the EEC to 'assure a common standard of health protection by adopting a rational set of scientifically based intervention levels for foodstuffs'. This is clearly an unattainable ideal and it should be realized that those who are asking for it are those who shrank back when it was offered. Unfortunately the position of the public in this political debate, for that is what it is, is difficult to judge. There is a growing concern, however, that health protection may come a poor second to trade protection.

Table 4.7 · Revised EEC Intervention Levels (all figures in Bq/kg)

Radionuclides	*Milk/infant food*	*Other major foods*	*Liquid food*
Strontium	125	750	100
Iodine	500	2,000	100
Plutonium etc.	20	80	10
All other nuclides of half life >10d	1,000	1,250	200

Summary

This chapter has primarily considered the Chernobyl incident and to some extent the Windscale and Three Mile Island accidents in relation to government reactions. The plethora of restriction levels in different countries at different times can have done no more than baffle the public and increase their anxiety. Table 4.8 brings together the varying edicts that have issued from the EEC since Chernobyl and have been discussed in this chapter. A glance at this table will emphasize more than anything the need for a credible unified approach. However, it is clear that trade and political expediency have often been placed ahead of science and safety considerations. Nevertheless, it must be said that this, almost by chance, has generally produced more restrictive action levels than could have been justified by science alone. There is, however, obviously a spectrum even in this level of (apparent) overprotection. The point is that unless intervention or action are standardized, there will be conflict.

In future nuclear accidents in Europe, the maximum permitted levels of, for example, radiocaesium contamination in major foodstuffs, is to be 1250 Bq/kg, which is 25 per cent higher (less restrictive) than the level in force in the UK now (1989). This is at a time when the NRPB has perceived that total radiation risk rates are nearly five times higher than existed when the action level was derived. Since that time, the NRPB has also recommended that the maximum dose to the public from one site should not exceed 0.5 mSv per year. This is in fact one-tenth of the *legal* public dose limit in the UK. Inconsistencies like this reduce the credibility of the regulators in the same way as the lack of information given to the public after the 1957 Windscale fire did. We seem to have learnt very little in thirty years.

Table 4.8 Development of EEC Intervention Limits for Foodstuffs in the Wake of the Chernobyl Accident

Intervention limit (Bq/kg or Bq/litre)	*Commodity*			
	Dairy produce	*Other produce*	*Drinking water*	*Animal feeds*
Immediate intervention limits introduced under regulation by Commission, 6 May 1986, for:				
I-131	500	350		
Temporary intervention limits recommended by 'Article 31 Expert Group', May 1986 for:				
Cs-134 and 137	1,000	1,000		
Temporary intervention limits proposed by Commission and adopted by Council, 30 May 1986, for:				
Cs-134 and 137	370	600		
Permanent intervention limits recommended by 'Article 31 Expert Group', Sept. 1987, for:				
I-131 and Sr-90				
year 1	700	7,000	500	
> year 1	500	3,000	400	
Pu-239				
year 1	80	400	60	
> year 1	20	80	10	
Cs-134 and 137				
year 1	20,000	30,000	3,000	
> year 1	4,000	5,000	700	
Permanent intervention limits under discussion in the Commission, Dec. 1986, for:				
I-131 and Sr-90	500	1,000	400	
Pu-239	20	30	10	
Cs-134 and 137	4,000	2,000	700	
Permanent intervention limits recommended by 'Article 31 Expert Group', April 1987, for:				
I-131 and Sr-90	500	3,000	400	
Pu-239	20	80	10	
Cs-134 and 137	4,000	6,000	800	12,000
Permanent intervention limits proposed by Commission to Council, May 1987, for:				
I-131, Sr-90 and Pu-239	as expert recommendations above			
Cs-134 and 137	1,000	1,250	800	2,500

5 Leukaemia 'Clusters'

The human species has survived and developed in a wide variety of natural environments. In the last few decades, however, man's activities have begun to pollute his environment severely, creating physical and chemical tests for health on a global scale that may eventually threaten human survival. Nevertheless, we have been complacent because the net effect of industrial progress has been beneficial – people live longer and are generally healthier. The world population continues to spiral inexorably upwards and there is consequently an ever increasing need for energy production. At almost the eleventh hour we have started to consider the local and global effects of this thirst for power. Although, the nuclear power industry has yet to make a large global impact on energy production (see Figure 5.1) it features strongly in people's minds as a potential source of environmental pollution. As an example of this, the concern over small 'pockets' or 'clusters' of leukaemia near nuclear sites may yet be seen as a serious brake on the expansion of the nuclear power industry in the UK.

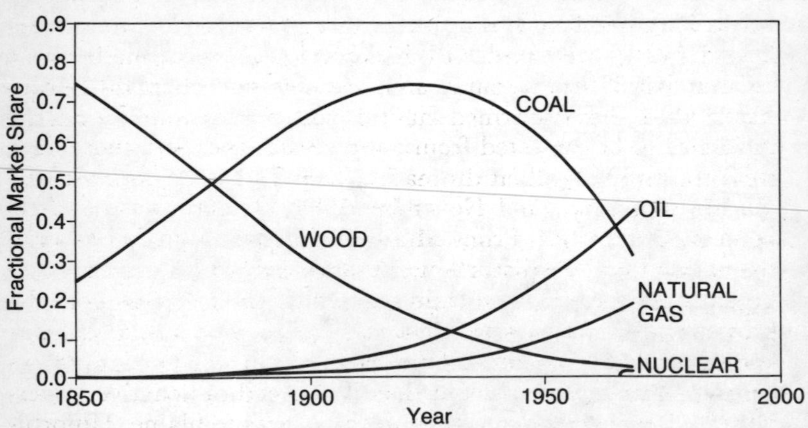

Figure 5.1 The History of Global Primary Energy Substitution.

The identification and recording of small pockets of rare diseases like leukaemia goes back well before the beginning of the use of nuclear power. However, the attention of the public was first drawn to it in 1983 by James Cutler, an investigative TV producer. Cutler, of Yorkshire TV, had earlier produced a brilliantly effective documentary on the effects of asbestos called 'Alice: a Fight for Life'. This film was concerned with the health effects, notably the incidence of a cancer called mesothelioma, in factories. In an effort to produce another health and safety film, he had approached the management of British Nuclear Fuels (BNFL) at Sellafield. Initially at least, Cutler was interested in the workforce at Sellafield (or Windscale as it was then called) even though there was little evidence from epidemiological studies carried out by BNFL that there were health effect problems.

In 1983, however, the National Radiological Protection Board had published a report on the health effects of the 1957 Windscale Fire that had revealed estimates of thirty-three cancer deaths from the incident. Journalists, including some from YTV, had flocked to Cumbria to investigate this and what they found surprised them. Cutler's team realized after talking to local residents that they should look outside the factory for health effects, notably childhood leukaemia in the area immediately to the south of Sellafield centred on the village of Seascale (population about 3000). They took their suspicions to John Urquhart, a statistician from Newcastle University, for scientific endorsement. They had by this time established that seven children under 25 years of age (five under 10) had developed leukaemia between 1954 and 1983 in this small area near the nuclear plant. Urquhart's researches confirmed that this was more than ten times the incidence to be expected from national statistics and these data were transmitted, albeit dramatically on TV, in 'Windscale: the Nuclear Laundry' on 1 November 1983. Quickly (on the afternoon of 1 November) Prime Minister Margaret Thatcher assured the nation that 'the disturbing claims ... would be properly and urgently examined'. The claims to which she referred were the programme's conclusions that, as exposure to radiation was known to cause leukaemia, if the disease occurred at a higher than expected rate near to a plant that discharged radioactive materials into the environment, then the plant was to blame. Unfortunately, nothing is as simple as that.

Initial reactions to the programme were predictable. The

consultant paediatrician responsible for childhood cancer registration for the whole of the Northern Region wrote to *The Lancet* with a bland denial of the existence of the problem, commenting that:

> There is no indication that the incidence of childhood cancer in Cumbria or on the Fylde coast is increasing or that there is an excess of childhood cancer in Cumbria or any of its subdistricts when compared to the rest of the north and north-west of England . . . The current publicity has generated a great deal of alarm and concern amongst the people of West Cumbria and we hope the publication of these figures will go some way to allaying their fears.

As pointed out by James Cutler in his subsequent book, the error in these conclusions was one of looking at too large an area and thereby diluting the effect.

Governmental reaction to the problem, although speedy and sensible, was typically British. A committee of inquiry was set up under the chairmanship of Sir Douglas Black, a past president of the Royal College of Physicians. However, the reaction of others was one of no surprise. Although the size of the excess incidence of leukaemia was unexpected, it had been anticipated as an epidemiological finding since the Windscale Inquiry in 1977. This public inquiry had been set up to consider the BNFL's application to build a reprocessing plant for oxide fuel, the so called THORP plant. Planning permission was eventually granted for this plant but the inspector, Mr Justice Parker, had suggested that more monitoring should be done on the Cumbrian coast. At the time of the Windscale Inquiry, it had come to light that dust in houses on the Ravensglass estuary, south of Seascale, was contaminated with plutonium and other actinides to a level between 100 and 10,000 times the level elsewhere in the country. Clearly this sort of contamination, if typical of the Cumbrian coast, might provide the environment in which an increase in cancer could be expected. Once again, things were not as simple as that, but the situation did, and does undeniably give cause for concern.

About three years after the Windscale Inquiry, Cumbria County Council initiated its own investigation into cancer levels in Cumbria and the North West. It appointed a community medicine general practitioner who, together with the medical officer for the West Cumbria health authority, eventually concluded that the Cumbrian cancer rate, taking all ages together, was slightly lower than the national average in the period 1951–78. This was exactly what BNFL wanted to hear, but shortly afterwards the YTV programme pointed to the flaws in such studies and it was left to the Black Committee to provide an answer.

The Black Committee

The Black Committee reported its findings in 1984. Apart from ten recommendations, it made a number of statements, some of which are worth recalling:

> The hypothesis in the television programme that the proximity of Sellafield to the village of Seascale could be a factor in producing cases of childhood leukaemia is not one which can be categorically dismissed nor on the other hand is it easy to prove.
>
> In the Northern Children's Cancer Registry region, which contains 765 wards, Seascale had the highest lymphoid malignancy rate during 1968–82 in children under 15 years of age in one study.
>
> However, the risks to the public of the Sellafield operation should not be judged against the standard of 'total or absolute safety', which is quite unattainable in any human activity . . . The fact that no operation can be made absolutely safe does not conflict with the desirability, indeed the necessity, of making it as safe as is practicable.
>
> It is impossible to establish for certain the situation with regard to environmental levels of radiation around Sellafield twenty or thirty years ago, and we shall never know the actual doses received by those children subsequently contracting leukaemia. In addition, one cannot completely exclude the possibility of unplanned discharges which were not detected by the monitoring programmes and yet delivered a significant dose to humans via an unsuspected route.
>
> We have no evidence of any general risks to health for children or adults living near Sellafield when compared to the rest of Cumbria and we can give a qualified reassurance to the people concerned about a possible health hazard in the neigh-

> bourhood of Sellafield. However, there are uncertainties concerning the operation of the plant which were highlighted in the Nuclear Installations Inspectorate Report of the November 1983 incident [the 'beach contamination incident'] and also problems attendant on the functioning of a plant, part of which has been long in service. There are further questions concerning the adequacy of the controls over present permitted levels for discharges; the quantitative assessment of apparent excesses of cancer, and possible genetic risks. During our investigations we also found some evidence of lack of co-ordination between the various agencies with an interest in this industry and considering its impact on the health of the community.

The Black Committee made five recommendations for further epidemiological work, four concerning the health implications of discharged radionuclides and one concerning regulatory mechanisms. As a direct result of this last recommendation, the DHSS set up a committee to review the effects of radiation on the community (Committee on the Medical Aspects of Radiation in the Environment, COMARE).

The Black Committee was very cautious in its appraisal of the data presented to it. The case was not as clear-cut as it had appeared at first sight, but most importantly, it was evident that contemporary radiobiological knowledge was insufficient to give a simple explanation. Since the initial 'cluster' of leukaemia was investigated, a number of others have been reported; in fact there now seems to be an increased incidence or cluster of cancers around all nuclear sites in the UK irrespective of the level of discharge.

The Black Committee, somewhat at variance with reports from other epidemiological groups, identified only seven cases of childhood leukaemia resident in Seascale since 1955 and under 25 years at diagnosis. Two of these children were born outside the area (Millom Rural District, see Figure 5.2) and of the rest, one was still alive. Thus, four fatal cases of leukaemia (two with chronic myeloid, one with chronic lymphocytic and one with acute lymphocytic leukaemia) were considered for comparison with national statistics and similar sized areas from elsewhere in the UK. The Black Committee focused its attention on the age group up to 25 years of age because that was the age group which had experienced an increased risk according to YTV. Since that

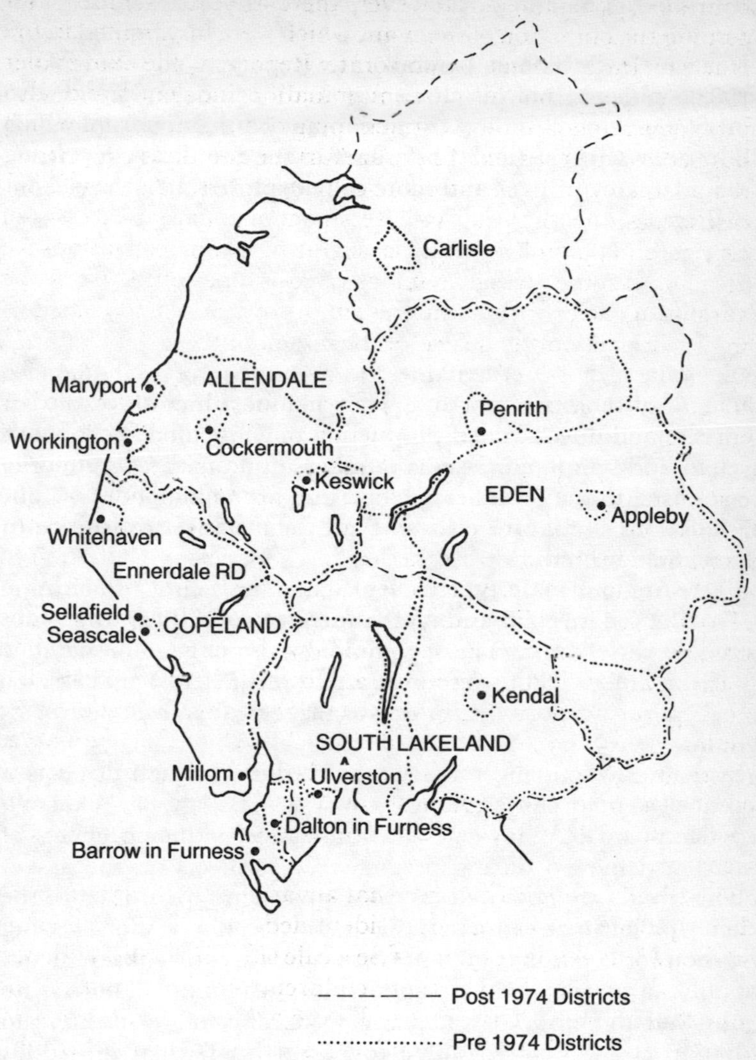

Figure 5.2 Cumbrian Rural Districts.

time, however, most epidemiological studies have looked at groups up to 15 years of age because that is taken as the upper limit of childhood and by then the incidence of leukaemia is at a minimum.

It is worthwile looking carefully at the epidemiology of leukaemia, the phenomenon of clustering and radiation risk rates to identify just how close we are to explaining the problems. This will be done with particular reference to the Sellafield experience as the data are stronger and more fully explored than those from elsewhere.

Leukaemia

Leukaemia is a generic name for a number of malignancies (cancers) of the bone marrow. They manifest themselves by an increased and uncontrolled production of white blood cells, both myeloid and lymphoid. Excess numbers of immature white cells are released to the peripheral blood and are unable to carry out the functions of mature cells so that, for instance, resistance to infection is lowered.

There are four main types of leukaemia, i.e. acute and chronic versions of the myeloid and lymphatic forms. Of these, the acute lymphoid variety occurs most commonly in young children and is the main cancer in this age group. However, leukaemia is rare overall, being responsible for only 1 in 40 deaths from cancer in Britain.

Exposure to ionizing radiation at medium to high doses is a well-documented causative agent in the aetiology of leukaemia (see Chapter 3). At lower doses, the cancer-inducing effects of radiation delivered during pregnancy, i.e. effects on the as yet unborn child, are now well established and are quantified to the extent that it can be estimated that natural radiation alone during gestation could account for 8 per cent of all cases of leukaemia and possibly, according to some authorities, even more. There is no doubt that the embryo/foetus is considerably more sensitive to radiation (by a factor of three to five at least) than the adult. There is, however, considerable controversy over the risk rate for leukaemia induction (see Chapter 3) but this was offset by the Black Committee in their choice of a factor that implied that *all* childhood leukaemia was caused by radiation (see later).

Some small contribution to leukaemia induction probably comes from genetic predisposition. This conclusion is to some extent speculative in that only in cases of recognizable hereditary abnormality, mostly chromosomal, has there been an increased incidence of childhood leukaemia. For example, there is a vastly increased (greater than twenty-fold) leukaemia risk in cases of Down's syndrome, which is a hereditary disease characterized by an extra chromosome 21.

There are a number of other factors that have been directly or indirectly cited as being involved in leukaemia. These include exposure to certain chemicals such as benzene, or treatment with chemotherapeutic agents for cancer or organ transplantation. Viral infection is also a likely cause because certain viruses are linked with leukaemia induction in mice and the HTLV-1 virus is known to be associated with adult lymphoma. Overall, the evidence of transmission of leukaemia from one individual to another is really minimal; if this phenomenon could occur, small epidemics would be expected from time to time.

As with other childhood cancer (e.g. brain tumours) the incidence of leukaemia, particularly the actute lymphatic form, increases rapidly after birth, reaches a peak at about age 3, and then declines. Because of the suspected environmental causes, the incidence might be expected to vary with country of birth. This is in fact so (see Table 5.1). For instance, in Europe, the rate varies between 31 and 49 per million, but much of this variation may be due to the relative efficiency of registration. Unlike other cancers, the variation of incidence with socio-economic group appears to be small, but this is a point of some controversy.

More obvious is the variation in trends of mortality over time. This increased fourfold between 1911–15 and 1951–5 and has halved in the last twenty years, reflecting the impact of more effective treatment. This changing mortality rate has important implications for the estimation of excess deaths from the disease. The rapidly increasing mortality in the years between the two world wars is of great interest. The most likely explanation is that, as infections are common in the early stages of leukaemia, many children would have died from an infection (before the introduction of sulphonamides and antibiotics) before their leukaemias were diagnosed.

Table 5.1 Childhood Leukaemia Throughout the World (annual incidence per million children)*

Costa Rica	59
Los Angeles (Hispanic), Australia (NSW)	50
Italy	49
Spain	47
Norway	45
Los Angeles (other white), New York (white), Hong Kong, New Zealand (non-Maori)	45–9
Switzerland, Sweden	44
Federal Republic of Germany	43
Finland	42
USA (white), Canada (western provinces), Brazil (Fortaleza), Puerto Rico, China (Shanghai), Japan, Philippines, Colombia (Cali)	40–4
France	40
Denmark, Czechoslovakia	39
Canada (Atlantic provinces), Singapore, Australia (Queensland)	35–9
England, Wales, Scotland, Netherlands	38
German Democratic Republic, Hungary, Yugoslavia	34
Poland	31
Brazil (São Paolo), Cuba, Israel (Jews), New Zealand (Maori)	30–4
USA (black), Los Angeles (black), New York (black), Brazil (Recife), China (Taipei), Israel (non-Jews),	25–9
Jamaica, India (Bombay)	20 4
Fiji	18
Zimbabwe (Bulawayo)	16
Uganda (Kampala)	14
Kuwait, India (Bangalore)	13
Nigeria (Ibadan)	12

Clustering and Statistics

The clustering or increased incidence of rare diseases in both space and time has been recorded since registration of diseases and the science of epidemiology developed. Even with childhood leukaemia there were reports of 'clusters' long before the so-called nuclear age began. So the increased mortality at Seascale does not appear to be a new phenomenon, but is it therefore to be expected merely by chance? For the answer to this question, we must investigate the statistics of Poisson probabilities.

Statistical fluctuations in the incidence of rare diseases always occur, because of the small numbers involved; only four cases of childhood leukaemia seem to have occurred in Seascale in twenty-five years. It is also noteworthy that COMARE, in its reappraisal of the evidence presented to the Black Committee, found an additional case of leukaemia (in a child who had moved from the area). This emphasizes the difficulty and importance of ascertaining the complete data set when numbers are so small. However, as such fluctuations can occur by chance, how can we be sure that chance was not operating in Seascale? It is common practice in such cases to calculate the probability of the occurrence of, say, a cluster of leukaemias by chance using the Poisson distribution (this is a distribution describing the frequency of random events each with a uniform probability). The assumption here is that the events (cases) were in fact random and not associated, which is a reasonable assumption. The Poisson distribution can be used to predict how many cases might be expected in certain-sized populations. If, for example, an average of three deaths from leukaemia were to be expected in any community of 4000 people of all ages during a given period of time, or estimated from national statistics, chance alone would cause this number of three to vary from zero to twelve if communities of equal size were examined throughout the land. In this example, in a total population of 56 million there would be 14,000 communities of 4000 people. Of these 470 of them would have 7 or more cases, 15 would have 10 or more and 1 would have 12 or more (i.e. four times the national average).

In addition, by extending this sort of calculation to larger and larger populations it can be shown that the effect may be obscured or diluted. For instance, Table 5.2 shows that as the population increases, the ratio of the maximum number of cases in the Poisson distribution to the average decreases.

In Seascale, the four cases originally noted exceeded expectations from national statistics by more than a factor of ten. The possibility of this occurring by chance was therefore very small indeed (about 1 in 8000). In fact in the 765 electoral wards of

Table 5.2 Ratio of the Maximum Number of Cases in the Poisson Distribution to the Average According to Population Size

Size of each subpopulation	*Average number of subpopulation*	*Maximum number of cases in any one subpopulation*	*Ratio max no. to average no.*
4,000	3	>12	>4
8,000	6	>17	>2.8
16,000	12	>26	>2.2
40,000	30	>49	>1.6

North Region examined by the Black Committee, Seascale ranked sixth highest for incidence rates for all childhood lymphoid malignancy but easily first using the Poisson criterion; it had lowest probability that the observed number of cases had occurred by chance. This probability is normally expressed as a Poisson probability or *p* value. A value of *p* of 0.02, i.e. 1 in 20, is considered to be of significance, meaning that there was a less than 1 in 20 chance of the occurrence (of whatever is being tested) being purely by chance. The *p* value of the leukaemia incidence rate in Seascale was 0.000134, i.e. a less than a 1 in 10,000 chance occurrence.

Nevertheless, it has to be said that even this level is not complete certainty, and when the Black Committee looked at the geographical distribution of those electoral wards with the highest incidence of lymphoid malignancy in Cumbria, there was no obvious connection with the Sellafield site or even the west coast. The Committee was also keen to point out that although the observed incidence of leukaemia was sixteen times that expected, this was only represented by four cases. From the evidence presented to it, the Black Committee took the view that the incidence of leukaemia in Seascale needed some explanation and for this they looked to the NRPB.

Radiation Exposure of the Children of West Cumbria

As the original YTV programme had suggested a connection between radiation exposure of children (notably those under 25) and leukaemia, the Black Committee asked the NRPB to do a radiological assessment of that group. At first sight it seems reasonable to assume a connection, but in reality the causal relationship is much more obscure. The NRPB approached the task in a scientifically responsible manner and produced a 300-page report in July 1984 followed two years later by a 158-page addendum when additional Sellafield discharges came to light.

From the outset the aim of its work was to estimate the leukaemogenically effective radiation dose to the Seascale children. These radiation doses came from four sources:

(a) releases of radioactive materials from Sellafield to the sea and to the atmosphere both from routine operations and also from incidents and accidents;
(b) natural background;
(c) fallout from weapons testing; and
(d) medical exposure.

The radiation doses from both external radiation and from intakes of radioactive materials were either estimated from measured values or calculated from discharge data. These data were used to calculate risks of leukaemia and other fatal cancers for seven cohorts of children assumed to have been born at five-yearly intervals from 1945 to 1975. Each of these seven cohorts comprised 175 children, which is approximately the average number of children born in Seascale in each five-year period over thirty-five years (1225 children were actually born in this period).

Clearly, none of the doses could be estimated very precisely. Even if the actual levels in the environment had been known for all of the thirty-five years, the intakes were not, and even the anatomical location of the critical cells at risk in the baby is still uncertain. In this absence of sound information, the NRPB had to resort to the application of models and assumptions. Predictably, these were attacked after publication of its report but the report did expose some of the deficiencies in control and monitoring that had existed in earlier years. For example, intakes of plutonium-239 and americium-241 before 1977 had to be estimated from reported discharges, as monitoring for these isotopes was, incredibly, not introduced until that year. This also applied to plutonium-241 before 1981. Another uncertainty was the value for the fraction of plutonium absorbed from the gastrointestinal tract. As the dose from plutonium and the other actinides was seen to be of some importance and young children apparently tend to ingest more house dust, sand, etc. than adults, this value was critical. At the time, the ICRP was recommending a value of 0.01 per cent, but the NRPB used a value of 0.05 per cent for plutonium complexed in food. Subsequently ICRP increased its value to 0.1 per cent but other reviewers have recommended the use of values of 0.3–0.5 per cent for young children. The NRPB was also criticized for possibly underestimating the dose from polonium-210 released during the Windscale fire in 1957. Again, because of a dearth of information, the NRPB had made value judgements over air concentrations and lung retention of this radionuclide that could have underestimated, so their critics claimed, dose by up to a factor of ten.

Overall, in combination, the underestimates of the NRPB's dose estimate could be as great as 30–100. Bearing in mind also that the effect of alpha-emitting radionuclides on the embryo/foetus/neonate is still largely unquantified, there is good reason to have doubts about the NRPB's risk estimates.

However, in order to offset some of these doubts, the Black Committee assumed that background radiation caused *all* childhood leukaemias. This equates to a risk rate of 22.8 in 1000 per sievert of exposure, and is put into context in Table 5.3. Clearly, the risk rate used by the NRPB was quite conservative but it still left the possibility of an underestimate of dose (and therefore risk) that was worth considering.

Using these risk rates, the NRPB estimated that a maximum of

Table 5.3 Leukaemia Risk Rates

		Upper bounds (assumed by NRPB)
ICRP (1977) for adults	2.0 in 1,000 per Sv	
In utero	12.5 in 1,000 per Sv	230 in 1,000 per Sv
0–10 years	5.0 in 1,000 per Sv	26 in 1,000 per Sv
>10 years	3.5 in 1,000 per Sv	14 in 1,000 per Sv
All ages (in utero to 20 years) Black Committee	22.8 in 1,000 per Sv	
UNSCEAR 1988 (in utero)	25.0 in 1,000 per Sv	

0.1 radiation-induced deaths from leukaemia could have been expected to occur in Seascale between 1950 and 1980. Only 10 per cent of this could be ascribed to the Sellafield discharges (see Table 5.4) so that even if the risk associated with Sellafield had been underestimated by an order of magnitude (ten times), the predicted number of radiation-induced leukaemias would only be doubled. Based on national statistics, the number of leukaemias expected from all causes for all cohorts combined was calculated to be 0.5.

The NRPB's advice to the Black Committee proved not to be the end of the matter because in November 1984 the NRPB was approached by Dr D. Jakeman, an employee of UKAEA, who had been abroad during the time the Black Committee was collecting evidence. He produced evidence that a release of uranium oxide from Sellafield in 1954 had been much bigger than had been admitted by UKAEA. A 1955 report had apparently stated that 100–200 g of uranium oxide had been released; this figure was subsequently revised upwards in 1957 to 440 g. Jakeman gave evidence that revised this figure up to 20 kg. This evidence was given to the NRPB and it revised its calculations, producing the 158-page addendum already referred to. The numbers of leukaemias to be expected from Sellafield releases was revised upwards to 0.016 (see Table 5.5).

By this time, however, the role of medical watchdog had been taken over by COMARE, a committee set up by the DHSS. This committee's first report was on the reassessment of dose at Seascale and its implications. Apart from science, it commented strongly on the way in which the new evidence had come to light

Table 5.4 Contribution of Different Sources to Total Risk of Radiation-induced Leukaemia for Persons to Age 20 or 1980, in Each Cohort

	Percentage contribution for each cohort							
Source	*1945*	*1950*	*1955*	*1960*	*1965*	*1970*	*1975*	*Total*
Sellafield routine discharges	7.3	11.0	10.3	5.5	8.1	13.0	16.2	9.1
Windscale fire	1.3	3.5	2.9	0.3	0.2	0.2	0.1	1.6
Fallout	1.6	5.3	11.9	16.7	12.5	5.6	2.9	9.2
Medical	12.1	10.9	10.1	10.5	8.9	7.1	6.6	10.1
Natural background	77.7	69.5	64.8	67.0	70.3	74.0	74.1	70.1

Table 5.5 Risk of Radiation-induced Leukaemia in the Seascale Study Population, by Source

Source	*Predicted number of radiation-induced fatal leukaemias in study population*	*Contribution to total risk (%)*
Sellafield discharges	$1.4\ 10^{-2}$	14.2
Windscale fire	$1.8\ 10^{-3}$	1.8
Weapons fallout	$9.2\ 10^{-3}$	9.2
Medical	$9.1\ 10^{-3}$	9.1
Natural background	$6.6\ 10^{-2}$	65.7
Total	$1.0\ 10^{-1}$	100

Note Total radiation induced leukaemias from Sellafield releases = 0.014 + 0.0018 = 0.0158 (approx. 0.016).

and also said: 'We therefore consider that the level of uncertainty about the information available and about risk to the population from the Sellafield discharges is now greater than at the time of publication of the Black Report'. While the NRPB was careful to point out that the discharges from Sellafield were 250 times too low to account for the leukaemias observed, the DHSS committee was issuing cautionary reports. There must always be a worry in these circumstances that we do not know what has been discharged and whether some unsuspected pathway to man has taken precedence, especially when monitoring of the environment had been shown to have been so inadequate in the past.

The public must be bewildered by the amazing inconsistency of government utilities and committees in this area. BNFL reassurances about improved levels of discharges in the future are seemingly hollow in comparison to unrecorded discharges in the past and COMARE's complaint about 'rudimentary monitoring' and the fact that 'we shall never know the actual average doses received by the population that the cases are drawn from'.

Despite these uncertainties, however, the Seascale cluster of leukaemias was soon shown to be far from unique. By 1986, two events had occurred that must have further reduced the public's confidence in the nuclear industry. The first was the discovery of unrecorded discharges from Sellafield referred to above, the second was the publication of a report by the Information and Statistics Division (ISD) of the Common Services Agency for the

Scottish Health Service. This reported a marked excess of leukaemias, about twice that expected, among young children resident near the Dounreay nuclear establishment.

The Dounreay establishment, situated on the remote north coast of Caithness in Scotland (Figure 5.3), had begun its operations in 1958 with the start-up of the Dounreay materials testing reactor. The UK Atomic Energy Authority, which runs the site, employed an original staff of about 2250 people, of whom about 54 per cent lived in the nearest town, Thurso (population about 11,000). The establishment houses the fast breeder reactors, both the earlier Dounreay fast reactor (DFR) and the later prototype reactor (PFR). The site also includes the Royal Navy's experimental reactor, Vulcan. However, the site has some similarities with Sellafield as there is a small-scale nuclear fuel reprocessing facility that started operations in 1958.

In 1986 a public inquiry was held to consider a planning application for a new reprocessing plant at Dounreay. In preparation for this, and bearing in mind the Sellafield experience, it was considered advisable to carry out an epidemiological study. The study looked at the incidence of leukaemia in 0–24-year-olds in the period 1968–84. The choice of this period was slightly constrained in that although mortality data existed before 1968, cancer registration data (i.e. incidence) were only available from 1968. Also, the postcode location of each registration was only available from 1968 on the Scottish National Cancer Registry computer. The original survey considered leukaemia registrations in three areas – less than 12.5 km from Dounreay, between 12.5 km and 25 km, and greater than 25 km from the site – all within the Kirkwall postcode area of northeast Scotland. This geographical delineation was to present problems because the 12.5 km circle area neatly bisected the town of Thurso.

Twelve cases of leukaemia were identified in the period 1968–84, six in the area more than 25 km from the site, and five in the inner circle (all of which were registered as lymphoid or thought to be lymphoid in origin). More importantly from the point of view of the subsequent radiological assessment, four of these cases were resident in Thurso. All six of the cases living within 25 km of Dounreay had been registered since 1980.

With the experience of Sellafield in mind, COMARE closely examined the pattern of the leukaemias and the diagnoses in terms of cell type. It calculated expected numbers of cases and

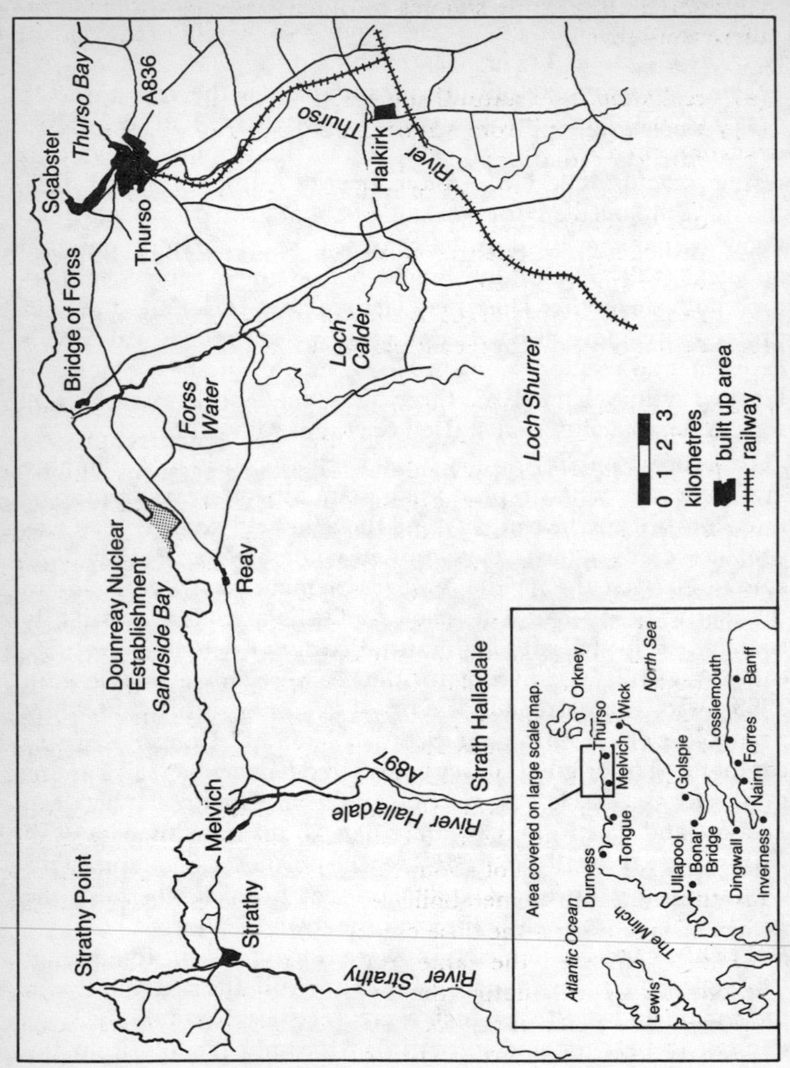

Figure 5.3 The Dounreay Area.

asked the NRPB to carry out an assessment of the radiation exposure of the children of Thurso. Once again, the NRPB considered all possible sources of radiation exposure. This time there were seven:

(a) radiation from natural sources;
(b) nuclear fallout from atmospheric testing of weapons;
(c) medical radiation;
(d) miscellaneous sources of radiation, e.g. colour TV, luminous watches, etc;
(e) radioactive discharges from the Dounreay site and from waste stored on site;
(f) radioactive discharges to the sea from Sellafield; and
(g) radioactive discharges from the Vulcan site.

The NRPB was again presented with what environmental monitoring information was available but, significantly, considered that it did not provide an adequate basis on which to assess the doses from discharges. The NRPB therefore based its assessment on discharge and mathematical models. This situation once again must be considered entirely unsatisfactory, partly because the NRPB admits that there are always uncertainties in environmental models, and partly because, once again, reliance was being put on the site operator's records, which had been found wanting at Sellafield. In addition, some of the NRPB's predictions of environmental levels were tested at the Dounreay Inquiry by an environmental pressure group, Friends of the Earth, and found to be inadequate. The comparisons of calculated to observed levels presented by the NRPB in their report seemed to indicate, with a few notable exceptions, underestimations by factors of about three. However, there is no doubt also that some of the metabolic models used would have had the effect of maximizing the dose calculated.

The NRPB used the same leukaemia risk rate factors as at Sellafield and calculated the expected number of radiation-induced leukaemias in the 4550 children born in Thurso between 1950 and 1984, as shown in Table 5.6. From this table it can be seen that the NRPB concluded that 0.34 cases of radiation-induced leukaemia could have been expected and that 0.005 might arise as a result of Dounreay and Sellafield discharges. However, as with Sellafield, this is only a very small proportion of the number observed. At Dounreay, there were four cases observed in Thurso. This was twice that expected from national

Table 5.6 Calculated Number of Radiation-induced Leukaemias in a Total of 4,550 Children Born in Thurso between 1950 and 1984, to Age 25 Years or to 1985, whichever is earlier

Source	*Number of radiation-induced leukaemias*	*Percentage contribution to total risk of radiation-leukaemia*
Dounreay discharges	0.004	1.2
Sellafield discharges	0.0009	0.3
Fallout	0.040	12.
Medical	0.026	7.5
Natural radiation	0.27	79.
Total	0.34	100.

Source NRPB R – 196

statistics although this increase was not really significant (p = 0.079).

In the COMARE report concerning this area attempts were made to explore possible errors in the estimates and assessments made. The committee was unable to offer any explanation for the increased incidence but pointed out that the radiation exposure was even less than in Cumbria. However, two significant points were made in the Committee's summary:

(a) It drew attention to a paper by some members of the Imperial Cancer Research Fund that pointed out that risks calculated from actinides, e.g. plutonium, are unlikely to be underestimates as evident changes in leukaemia incidence would have been associated with variations in fallout; these had not been observed either in the UK or elsewhere. This view was discounted by COMARE, 'leaving open the possibility that a particular form of a specific radionuclide, which emits high LET (e.g. alpha) radiation, may produce greater risks of leukaemia in young people than previously thought'.

(b) The Seascale cohort studies had indicated that the excess of leukaemias seemed to be concentrated in those children born to mothers resident in Seascale. No excess was seen in children born elsewhere who subsequently moved to the

> area. This suggested to COMARE that if radiation was a causative agent, it probably would have to act on the embryo/foetus (i.e. in pregnancy) or on the neonate. It was also evident that a majority of one of the parents of each of the leukaemic children had worked at Dounreay. The relevance of this information was not evaluated, but was thought to be of significance.

Overall, COMARE accepted that the increased incidences of lymphoid leukaemia near Sellafield and Dounreay were unlikely to have occurred by chance but could not be explained by radiation exposure using current knowledge of radiation biology. This is especially pertinent when the disparate calculated radiation estimated cases are considered, i.e. 0.016 out of 4 at Sellafield and 0.005 out of 4 at Dounreay. This implies gross underestimates of exposure that themselves differ by a factor of three.

The situation then became even more complex when more cancer 'clusters' were identified.

Leukaemia and Cancer Around Other Nuclear Sites

Looking for cancer 'clusters' soon became a nationwide occupation and increased incidences were easily found by both amateurs and professionals! In James Cutler's next documentary, 'Inside Britain's Bomb', broadcast in December 1985, he identified an area of increased cancer incidence near the Atomic Weapons Research Establishment at Aldermaston and near the Royal Ordnance Factory at Burghfield (see Figure 5.4). The childhood leukaemia aspect here had already arisen as a clinical suspicion at the Royal Berkshire Hospital in Reading. This suspicion that an unusually high number of cases of childhood leukaemia were being recorded in West Berkshire was followed up by a number of studies and caught the interest of COMARE. It found that it could not dispute the increased leukaemia instances but was again at a loss to explain them. It once again commented on the relative inadequacies of the environmental monitoring, e.g. it noted that although MAFF had conducted some monitoring around the sites, no environmental monitoring had been carried out by AWRE between 1960 and 1978. Once again, the increased incidence is unlikely to have occurred by chance alone and the NRPB has calculated doses much smaller than those for Thurso, i.e. with current knowledge radiation is even less likely to be the culprit. The implication is that the leukaemia incidences cannot be explained simply by assuming that the risk factor for radiation leukaemogenesis is too low, because it would have to be wrong by substantially different amounts for the three locations.

Most significantly, COMARE says:

> Plutonium exposure of parents or children may therefore be the common factor.
>
> We have therefore considered what factors the Sellafield, Dounreay, Aldermaston and Burghfield (and Harwell) sites have in common. Sellafield and Dounreay are both reprocessing sites with similar activities and (qualitatively) similar discharges. It should be noted that, although no reprocessing is done at Aldermaston, Burghfield or Harwell, plutonium is handled at all these sites.

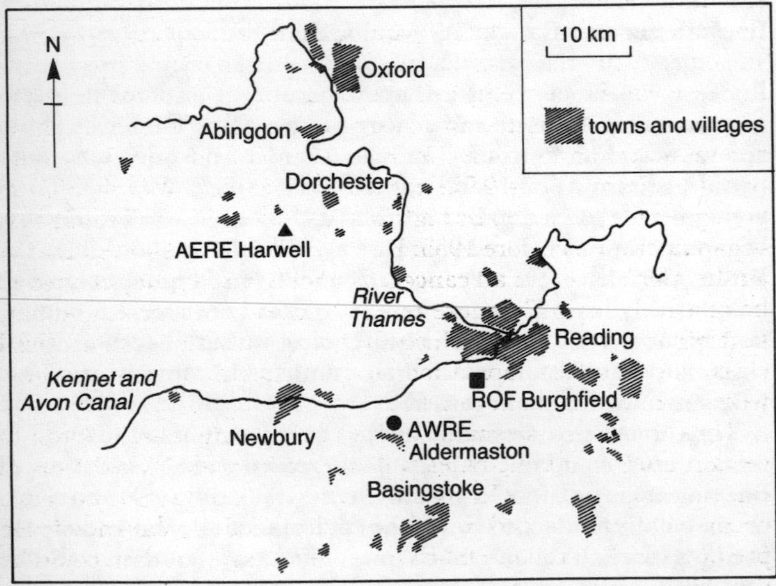

Figure 5.4 Nuclear Establishments in Berkshire.

However, increased incidences have now been reported near practically every nuclear site in Britain, most of which are nuclear power stations with extremely low discharge rates, notably Lydney near Berkeley/Oldbury, Leiston near Sizewell, the area around Hinkley Point and Ferndown near Winfrith. There are also reported increased incidences of other cancers around some non-nuclear sites emitting natural radioactive materials. These claims must all be taken seriously and have been studied extensively. Several recent (1989) papers including one by the Office of Population Censuses and Surveys (OPCS) have disregarded individual claims and normalized the studies to specific fixed distances from nuclear establishments, e.g. 10 miles or 12.5 km, in two fairly comprehensive papers. These voluminous data and results have been reported by the Imperial Cancer Research Fund which collaborated in the study. It looked at the incidence of cancer, particularly leukaemia, near all major nuclear establishments in England and Wales. Its analysis used data for pre-1974 local authority areas (LAA) with at least one-third of their population living within 10 miles of fifteen establishments. For each installation LAA, a non-installation LAA from the same standard region was selected as a control matched population through urban/rural status, population size, and socioeconomic structure. This report considers Sellafield separately (where the increase was 85 per cent) but mentions a significant increase (15 per cent) near Aldermaston. The results support the idea that in recent years the mortality from leukaemia and especially lymphoid leukaemia ($p = 0.01$) and Hodgkin's disease ($p = 0.05$) in young people tended to be high in areas close to installations that began operations before 1955. However, they also showed that in adults mortality from all cancers considered as a group tended to be relatively low. The more recent studies are especially important because variations with confounding variables such as social class, rural status, population size and health authority region were examined and allowed for.

Thus there now seems little doubt that within areas close to certain nuclear establishments there are increased incidences of certain cancers. These 'clusters' of cases are very unlikely to have occurred by chance and contemporary radiobiological knowledge predicts cancer at a very much lower level. So what then could be the cause?

Possible Explanations for Increased Incidences of Cancer

The first, most obvious, cause has already been discussed; the possibility that radiogenic risk rates are being underestimated. There is no doubt that the paucity of monitoring data, the validity of the environmental and metabolic models, and uncertainties in radiogenic risk rates for certain radioisotopes at early stages of life, e.g. in-utero, all conspire to make the prediction of risk a very uncertain process. More work is being done on all aspects of this problem. However, some of the increases identified are either not real or different causation processes must be operative because, as already described, there is a wide disparity between different sites in terms of the shortfall of predicted cancers. In addition, the doses from background radiation and from bomb fallout are, respectively, much greater and about the same as the doses from discharges. If risk rates or models are vast underestimates, different models would have to be operable for natural radionuclides and bomb fallout which, although plausible, is unlikely.

Another possibility is that the cancer cases are somehow genetically different. Genetically determined hypersensitivity to radiation is known, but is rare. However, it is possible for an undetected gene modification to be present in a subset of the population. Could this be responsible for the cancer excesses? It seems unlikely, mainly because empirical risk models are based on data from large population groups that would presumably include the genetically sensitive subset. Thus it would be necessary for the genetically sensitive groups to be present at Seascale and Thurso but not in Hiroshima and Nagasaki for instance. Also, of course, we come up against the problem of natural background which seems, assuming the models are correct, to be considerably larger than the dose from the discharges. The effects delivered by these two contributions to total dose would be unlikely to be altered by the presence of a sensitive subset.

There is also the possibility that, because of quirks of human behaviour, the intake of radioactive materials by some people is much higher than even the average members of the critical age groups. This is also unlikely because the increased intake would have to be so large as to make the anomalous behaviour pattern fairly obvious.

Although environmental monitoring in the past has been shown to be inadequate, it has to be hoped that it was good enough to detect large unrecorded discharges. However, it must be said that the increased discharges in the 1950s that were brought to the notice of COMARE were not suspected by the NRPB or the Black Committee members. Again, this *is* possible and when set alongside uncertainties in models, risk rates etc., it could be plausible as an explanation for underestimates of dose, at, say, Sellafield, but are the same things occurring at the other sites?

It has been suggested that the effects (rather than the exposure) might be indirect. For instance, the children are generally offspring of parents of whom one at least has been employed at the relevant nuclear installation. If the effect were genetic the dose to the child would be irrelevant, the dose to the parent would matter. There is little evidence of first-generation genetic effects in man but the low dose region is totally unexplored. This remains a first possibility as an explanation but the implications over the effects of natural background would need consideration.

One recent suggestion concerns the introduction of a leukaemic virus into isolated, remote communities by the influx of workers from other areas. It could not be used as an explanation for the increases of cancer found in Berkshire because this area could hardly be called an isolated community. Although the suggestion is interesting, the evidence supporting it has been criticized and at present it is no more than an interesting theory.

COMARE in its second report suggested that the increased incidence of leukaemia seemed to be associated with the mere presence of the nuclear establishment (it was referring to Sellafield and Dounreay). The suggestion that some other factor characteristic of the nuclear industry, even stress, may be responsible is difficult to explore, but case-control studies are being set up.

In several of the later, national epidemiological studies, adjustments have been made for socioeconomic and demographic factors. These have been effective and appropriate because, in adults, the relative risks of all malignancies and all nonmalignancies are close to one (0.97–1.07, typically). However, such adjustments do not reduce the relative risk of childhood leukaemia to one and there thus may be some other social factor that is important for children. In this respect, the tendency for leukaemia to occur in areas with relatively high proportions of their populations in socioeconomic groups I and II deserves further study. In Seascale, for instance, the proportion of the active male population in social class I was, in 1971, 47 per cent compared with 5 per cent nationally. However, no increased overall mortality has been observed in children born to parents in social classes I and II. Again this factor, if it exists, is being sought. Lastly, it would be instructive to compare epidemiological studies in the UK with studies in other countries. Unfortunately, there are few such studies.

International Evidence of Cancer Clusters

Evidence for increased incidences of cancers near nuclear establishments in other countries is variable. Information must of necessity depend on reliable cancer registration, and this has not always been available in other countries. However, there are a number of papers from the USA reporting excess cancer risks. The most closely studied site is the Rocky Flats plutonium facility near Denver, Colorado.

Plutonium weapons research and development has been carried out at this facility since 1953. There have been two substantial fires involving plutonium (in 1957 and 1969) that have produced contamination offsite. Carl Johnson, the then local Health Commissioner, has reported several times on an increased cancer and congenital malformation rate downwind from the site; he has also reported an increased cancer incidence among workers at the plant. Johnson's work on this population of nearly 600,000 people, which extends into Denver, has been criticized but there is now no question of the extent of plutonium contamination of the environment. There is, however, less certainty over the amount inhaled, and this confuses the interpretation. Also, lung tumour incidence is not a special feature of the study, which is a little odd. It has been reported that as a result of these studies Johnson was dismissed from his job as Medical Commissioner with Jefferson County.

Two other studies have shown no statistically significant increases in cancer incidence. One was carried out around the San Onfre power plant in California and the other, which is far more important, was undertaken around the French reprocessing plant at Cap de la Hague and reported in 1989. No studies from other countries approach the statistical significance of those conducted at Sellafield or Dounreay and none find excesses of leukaemias in children.

Lastly, one of the more intriguing associations claimed between radiation and health defects was reported in September 1986. In this report, it was noted that the fraction of the total annual number of deaths occurring in the USA in the months of May to August 1986 was higher than at any time since 1900. This was attributed to fall-out from Chernobyl, which reached the USA between the 7 and 10 May 1986. Moreover, the increased mortality was mostly concentrated in members of the 25–34 age group who were, the report noted, born at about the time of maximum fall-out from weapons testing in the USA. The report claimed that this cohort had a weakened immune surveillance as a result of their earlier exposure. It is difficult to associate Chernobyl with this effect, even if it is real, because the doses (of the order of 5 μSv) were so small compared with variations in natural background, etc. If the explanation offered is to be taken seriously, it is quite surprising this sub-group had lived to be 25. However, studies of this nature have to be considered before they are dismissed and are worthy of some attention.

Summary

There is clearly no easy way to explain away the increased incidence of childhood leukaemia at Seascale, and possibly the same applies to the situations near Dounreay and Aldermaston. The Black Committee and subsequently COMARE have expended time and effort in assessing the epidemiological findings and the possible radiation dose calculation. Money has been made available and appropriate research is being financed in attempts to throw light on areas where data were lacking, e.g. the long-term effects of high LET (alpha) radiation on the foetus. A committee has been set up to coordinate research financed jointly by BNFL, UKAEA and CEGB. All this means that the problem is recognized and taken seriously, but how far are we from identifying whether there is an association with discharges into the environment and long-term health effects?

In reality, the answer to that question depends on where we look. For instance recent reports suggest that exposure of the parent(s) of the children may be relevant, other reports indicate that in-utero exposure is important. It is noteworthy that these are areas in which there is little radiobiological knowledge and it is clear that current, conventional thinking is unable to assess a big enough dose or a big enough cancer risk rate to make the association.

It must be emphasized that if calculations of dose to the children are relevant, different models would be necessary at the three sites mentioned above (see Table 5.7). This is simply because the same increase in cancer incidence seems to come from areas of vastly different potential radiation exposure. Thus the expected number of cases of leukaemia calculated by COMARE in the 0–20 year age group from radioactive discharges was 0.013 per 1000 at Seascale, 0.00088 per 1000 at Thurso, and 0.000001 per 1000 at Aldermaston and Burghfield. However, increases in leukaemia have now been detected quite widely, particularly as a result of the data published by the OPCS in 1987 mentioned before.

COMARE has been certain that *some* factor associated with nuclear sites was causally connected with local high incidences of leukaemia; it now seems uncertain whether exposure of the parents or children is more important. It also now seems to realize that clusters of leukaemia occur remote from nuclear sites. However, its interest in plutonium and in-utero exposure seems at least to be appropriate. Nevertheless, the discrepancy between the dose required and the highest dose calculated leaves one wondering whether COMARE is clutching at straws.

It can be seen that the association between local increased incidences of cancer and radiation exposure, particularly childhood leukaemia, is still very much an open question. Nobody can be complacent enough to assert either that the pollution of the environment has nothing to do with this health effect or that the very existence of nuclear installations is the causative agent. Nevertheless the way forward is surely through a sober examination of all possible answers with adequately funded research and not through jumping to unconsidered conclusions prompted by emotional media reporting.

Table 5.7 Comparison of Dose Equivalents, μSv, to Red Bone Marrow of 1-year-old Children

Year	*In Seascale from BNFL, Sellafield*	*In Thurso from DNPDE, Caithness*	*In Thurso from BNFL, Sellafield*	*At 5 km from AWRE, Aldermaston*	*At 5 km from ROF, Burghfield*	*At 5 km from AERE Harwell*	*Weapons fallout*[1]	*Natural*[1]
1955	1,200	0	NC[2]	7.7×10^{-5}	0	0.38	54	990
1960	120	3.5	0.19	5.0×10^{-3}	0	0.61	90	990
1965	58	15.	0.37	4.9×10^{-3}	0	0.61	260	990
1970	120	8.9	0.94	1.5×10^{-2}	3.5×10^{-7}	1.6×10^{-2}	85	990
1975	330	3.1	4.8	2.4×10^{-2}	8.6×10^{-7}	1.6×10^{-2}	42	990
1980	170	0.30	4.2	1.2×10^{-2}	4.3×10^{-6}	1.8×10^{-2}	27	990

Notes

1 Values typical of Oxfordshire and Berkshire. Fallout doses in Thurso and Sellafield are up to a factor of two or so higher.

2 NC = not calculated.

6 Natural Background Radiation

Introduction

There is little controversy about the magnitude of natural background radiation, although recent reassessments have increased the doses by about 25 per cent. However, the doses received by everyone are often used as a standard by which to measure (or even justify) man-made radiation. This is a rather odd stance because we can do very little about natural background whereas we should be able to choose whether or not to accept additions to it. If it is to be used as a yardstick, though, it is worth considering where it comes from, its magnitude, how we are exposed to it, and whether we need to take measures to reduce it.

Natural background radiation is all around us because we are exposed both from above, through cosmic rays, and from the soil and rocks, through terrestrial radionuclides. At the time of the formation of the solar system, all the isotopes we know (and some more), both stable and unstable, were present and *Homo sapiens* has thus evolved in an ever-decreasing radiation field. In fact about 4,000 million years ago the background radiation dose would have been about twelve to fifteen times what it was when man first came on the scene. Thus we do not know what difference this radiation field has made to man's evolution. However, knowing what we do now about the biological effects of radiation, it seems likely that natural background is responsible for some ill-effects, such as cancers. Assuming this is so, we should view any addition to background with some circumspection. The alternative theory is that as we evolved in a rádiation field to which we apparently adapted, any radiation doses within the variations in this field should be tolerated. The former theory is more appealing.

Cosmogenic Radiation

Cosmic radiation refers not only to the primary energetic particles of extra-terrestrial origin that strike the earth's atmosphere but also to the secondary radiations produced by this interaction. It is this secondary radiation that impinges on human beings. Without the atmosphere we would be exposed to radiation doses at least a thousand times greater than those that exist on the surface of the earth. The origin of cosmic radiation is partly extra-galactic and to some extent of solar origin (the solar wind). Only at times of solar flares are the solar particles energetic enough to penetrate the atmosphere.

The secondary radiation that reaches the earth consists of a directly ionizing component and a small contribution from a neutron component. In the UK, the average outdoor dose rate from the directly ionizing component is about 280 μSv per year (range 260–290 μSv). Buildings afford some shelter and taking occupancy into account, the average annual dose from this component is about 230 μSv. The average neutron dose is about 3.5 μGy which, using a quality factor of six for neutrons of cosmic energies, represents a dose equivalent of 20 μSv per year. Thus, the total average annual dose equivalent from cosmic radiation is about 250 μSv (range 200–300 μSv) in the UK.

The higher the altitude at which one lives or works, the higher the dose. For instance, the dose doubles at 1800 m and triples at 3000 m. In the highest human domiciles (4000 m) in the Himalayas, the dose is six times the sea-level dose. At the altitude used by intercontinental passenger aircraft (12,000 m), the dose rate is 170 times the sea-level dose while for Concorde travellers the dose rate is higher still, by a factor of two. The 20,000 aircrew employed in the UK receive a collective dose of about 40 man-Sv and an average annual individual dose of 2 mSv. Concorde is fitted with radiation monitoring equipment and doses to its aircrew are higher at about 2.5 mSv per year, with a maximum of about 17 mSv. The dose to a trans-Atlantic traveller in Concorde, however, is about 20 per cent less than in a subsonic plane because of the difference in journey times.

Terrestrial Sources of Radiation

The rocks and soils of the earth contain primordial radionuclides that give rise to a radiation dose to the population. This comes partly from external radiation (gamma radiation) and partly from internal radiation as a result of eating or breathing the radioactive material. The majority of the radiation dose comes from radionu-

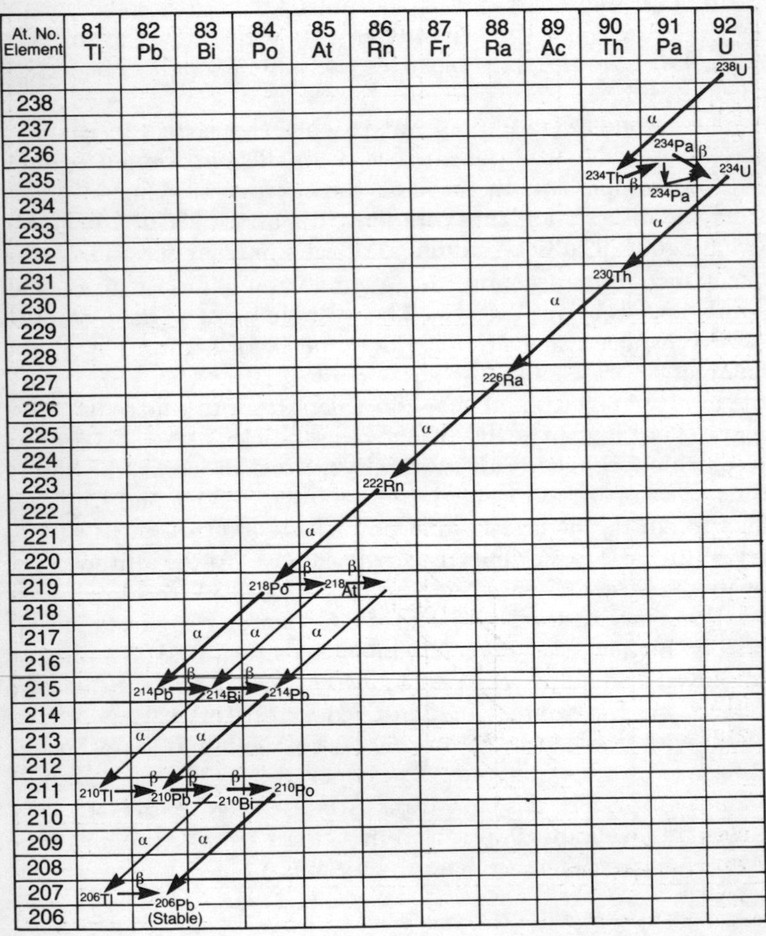

Figure 6.1 Uranium/Radium – (4n + 2) – Series.

clides in two of the three natural radioactive decay series. Two of these are illustrated in Figures 6.1 and 6.2. There are also a few other natural primordial radioisotopes that contribute to the public dose. Notable among these are potassium-40 and rubidium-87.

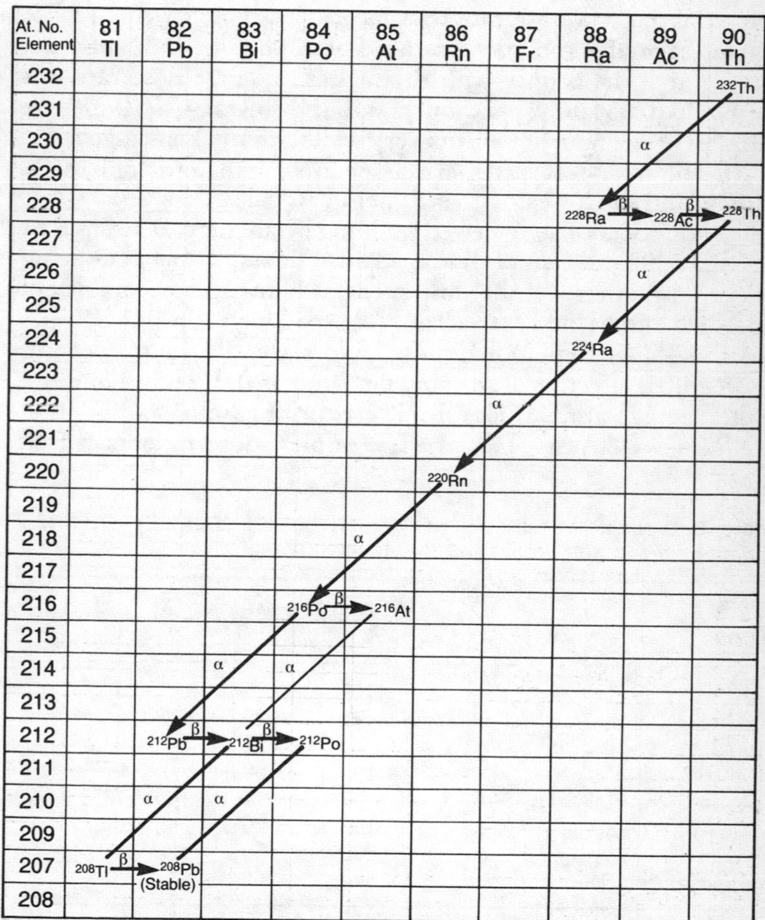

Figure 6.2 Thorium – (4n) – Series.

External gamma radiation

Some of the natural radionuclides in the ground, rocks, soil, etc. emit penetrating gamma radiation. In general, the contribution from the U-238 decay series, the Th-232 decay series and K-40 are 35 per cent, 25 per cent and 40 per cent respectively. The typical average concentrations in different media are shown in Table 6.1.

It can be seen that the highest levels are found in granitic rocks and this has been emphasized in the countrywide surveys of terrestrial gamma dose rate that have been done by the NRPB. The average absorbed dose rate was found to be 0.034 μSv per hour in air. The county with the highest average dose rate was Cornwall (0.054 μGy per hour) and the lowest was Berkshire (0.021 μGy per hour). This fits very well with the known geology, i.e. the highest dose rates are associated with hard old rocks, granites and the lowest with sedimentary rocks.

The NRPB has also reviewed building occupancy. It seems that on average 92 per cent of time is spent indoors, 77 per cent being spent in dwellings. Of the time spent at home, the occupancy of bedrooms and living areas accounts respectively for about 55 per cent and 45 per cent. When these factors are taken into account, the weighted effective dose equivalent rate is 0.042 μSv/hour and so the average annual dose from terrestrial gamma radiation in dwellings is 280 μSv. The average annual dose indoors (in all

Table 6.1 Typical Activities per Unit Mass of Naturally Occurring Radionuclides in Soil, Rocks and Building Materials

Material	*Activity* per unit mass (Bq kg^{-1})* ^{40}K	^{238}U	^{232}Th
Soil	370	25	25
Limestone	90	30	7
Sandstone	370	20	10
Granite	1,100	90	80
Sand and Gravel	30	4	7
Cement	150	20	20
Clay bricks	600	50	45
Lightweight blocks	370	60	25

Note *The values for uranium-238 and thorium-232 refer to the parent radionuclides. The total for all members of each decay series is variable and depends on the state of equilibrium throughout them.

buildings) is about 384 μSv and the overall dose from gamma rays is about 350 μSv per year. The range of individual doses is about a factor of two or three either side of the average, i.e. the range was 120–1200 μSv per year. A small fraction is above 0.1 μSv per hour (876 μSv per year) and some are below 0.02 μSv per hour; the highest averages were for Cornwall (500 μSv per year) and Devon (360 μSv per year). However, to these doses must be added the internal dose from radionuclides taken into the body, either by inhalation or ingestion in food.

Internal Exposures

Of those naturally occurring radioisotopes taken in by ingestion and inhalation, K-40, Rb-87 and radionuclides in the decay series of uranium and thorium contribute most. There is also a small contribution from radionuclides formed in the upper atmosphere by cosmic rays.

It is convenient to divide these 'internal emitters' into long-lived radionuclides, and radon and its daughters.

Long-lived radionuclides

Only three cosmogenic radionuclides are really of any significance to man (C-14, Be-7 and H-3) and only two of these (C-14 and Be-7) have any dosimetric impact. C-14, which is formed in the upper atmosphere from N-14, has some importance because it can be spread throughout the biosphere, i.e. plants, animals, food, etc. However, it is a low-energy beta emitter, and only gives rise to a mean annual individual dose of about 12 μSv. Interestingly, if no more man-made C-14 was released, the current level of C-14 labelled carbon dioxide in the atmosphere would be diluted by C-14 of lower specific activity (which had decayed through many half lives) being burnt and released from coal. The annual dose from Be-7 is quite small and averages about 3 μSv.

The average person contains about 4000 Bq of K-40, which is an isotope of potassium present in natural potassium to the extent of 0.0118 per cent. The amount in the body is controlled by metabolic reactions; for instance, the activity per unit mass is higher in young men. The average annual dose from K-40 is about 180 μSv.

The average annual doses from the longer-lived members of the uranium and thorium decay chains are given in Table 6.2. The values of dose vary markedly with diet but the major fraction comes from the Pb-210/Bi-210/Po-210 triad. The total mean annual dose from these longer-lived radionuclides is about 300 μSv, although these vary by about a factor of two to three about the average.

Radon and its daughters

In each of the three natural decay series of heavy elements, there is one link which is a radioisotope of a noble gas, viz:

U-238 .. Ra-226 Rn-222 ('radon') ..

U-235 .. Ra-223 Rn-219 ('actinon')

Th-232 .. Ra-224 Rn-220 ('thoron')

These three gases differ markedly in half life (for radon it is 3.8 days; for actinon, 3.9 seconds; and for thoron, 54 seconds).

Table 6.2 Average Annual Doses to the UK Population from Natural Sources of Internal Exposure*

Source of radiation	*Annual dose (μSv)*
Cosmogenic radionuclides	
^{7}Be	3
^{14}C	12
Primordial radionuclides	
^{40}K	180
^{87}Rb	6
^{238}U series:	
$^{238}U \rightarrow ^{234}U$	5
^{230}Th	7
^{226}Ra	3
$^{210}Pb \rightarrow ^{210}Po$	64
^{232}Th series:	
^{232}Th	3
$^{228}Ra \rightarrow ^{224}Ra$	13
Total (rounded)	300

Note *Excluding radon, thoron and short-lived decay products.

Because of this and the half lives of the subsequent chain, Rn-222 and the uranium decay series are the most important in terms of impact on man. We will therefore look at this chain in more detail.

If the series has been undisturbed by chemical and physical processes, there will be a radioactive equilibrium. This occurs when the series decays with a decay rate equal to that of the parent, U-238. Globally there may be equilibrium but locally processes tend to separate them into groups:

(a) U-238, Th-234, Pa-234m, U-234 and Th-230 tend to behave as a group. However, due to physical and chemical processes in the sea, Th-230 tends to precipitate preferentially and concentrates in silts and sediments. In the body, uranium accumulates slowly in bone, but in a soluble form it is considered as chemically toxic to the kidney.

(b) Ra-226 readily separates from the other members of the series and is distributed rather differently in the environment. It follows calcium and barium in its chemical reactions and metabolism. Thus more than 90 per cent of the body content of radium-226 is in the skeleton. Some of the bone dose comes from decay of its daughter, radon-222, and the latter's short-lived daughters, because only one-third of gaseous radon produced in the body is exhaled.

(c) Radon-222 is an inert gas that has no chemical compounds and does not really enter into any chemical reactions. As a gas, it may separate from its parent and enter the atmosphere. In a closed space it may be practically in equilibrium with its short-lived daughters, Po-218, Pb-214, Bi-214 and Po-214.

The alpha radiation from these short-lived daughters gives a significant dose to the lung (more strictly, the bronchial epithelium), while the radiations from Pb-214 and Bi-214 in soil give practically all of the terrestrial gamma dose from the uranium series. Radon is slightly soluble in body fluids (more so in fats) but the resulting doses are small when compared with those to the respiratory system.

(d) The long-lived daughters Pb-210, Bi-210 and Po-210, although initially in equilibrium, get separated by physical and chemical processes in the body. Lead (Pb) accumulates in the bone, and the daughter, Bi-210 decays in place. However, the fate of Po-210 is in some dispute because

> Po-210 is distributed in soft tissues as well as in bone. The primary source of Po-210 is Pb-210 decaying in the body, because the longer half life (22 years) of Pb-210 allows the build up of a much larger body content than the 138-day half life of Po-210. The significant bone dose comes from the alpha emission of Po-210, but is the marrow dose likewise affected? This is becoming a burning question in radiation biology nowadays as questions are asked about the leukaemogenic effect of natural radiation.

Uranium-238, at the head of the chain, has a half life (4.5×10^9 years) that is comparable to the age of the earth. Uranium in normal soil is present at a concentration of about 20 Bq/kg (range 7–40 Bq/kg).

Radium-226 tends to occur with about the same concentration as U-238. Wastes from uranium and phosphate mining contain significantly higher concentrations of radium, for instance, uranium tailings can have typically 100–500 times the concentration of normal soils, the extreme being about 14,000 times.

Radon escapes from soils relatively easily (75 per cent from the top 2 m), a typical concentration in air out of doors being 4 Bq per m^3. Daily and seasonal variations may, however, affect this by a factor of two or so. The radon concentration over material with a high radium content, e.g. near a uranium mine or just over granitic soil, can be as much as a hundred times normal values. In the UK high levels of radon in the air are a feature of Devon and Cornwall. The granitic rocks in these areas contain significant (1.5–8 per cent) quantities of uranium (and therefore radium). Several of the now disused tin and copper mines contain very significant concentrations of radon, and the amount in houses is now causing concern. The NRPB has carried out surveys of houses and found an average for Cornwall of 110 Bq per m^3 and for Devon of 74 Bq per m^3, compared to a countrywide average of 20.5 Bq per m^3. In the southwest of England as a whole, 20 per cent of the houses (13,000) had levels that exceeded 400 Bq per m^3 and 50,000 exceeded 200 Bq per m^3; the highest concentration found was 8000 Bq per m^3!

Despite previous concern over building materials, e.g. the high use of granite in Aberdeen, the ground underneath the house seems to be the greatest source of radon. This is particularly true where houses have been built on land-fill from mines, e.g.

uranium tailings in Colorado, shale tailings in Sweden, radium factory tailings in Australia and land reclaimed from phosphate mining in Florida. Radon concentrations in the upper storeys tend to be lower than on the ground floor. In Norway a survey showed that wooden houses had higher concentrations than brick ones. The thickness and integrity of the floor determines how much radon seeps in and therefore sealing the floor and walls also reduces the radon level; even wallpaper cuts down emissions. Once radon is there, of course, double glazing and draughtproofing make sure it stays inside the house. The NRPB estimates that these accompaniments to modern living increase radon concentrations by up to 30 per cent. Ironically, although a good wood- or coal-burning fire may increase carbon dioxide in the atmosphere, it reduces radon inside houses by providing ventilation.

Despite localized high levels, doses from exposure to radon in the UK are moderate when compared with some other countries, as Figure 6.3 indicates.

Dosimetry of Radon and its Daughters

After inhalation, radon and its daughters only irradiate the cells in the lung to any significant degree. There has, however, been considerable discussion about which cells in the lung are irra-

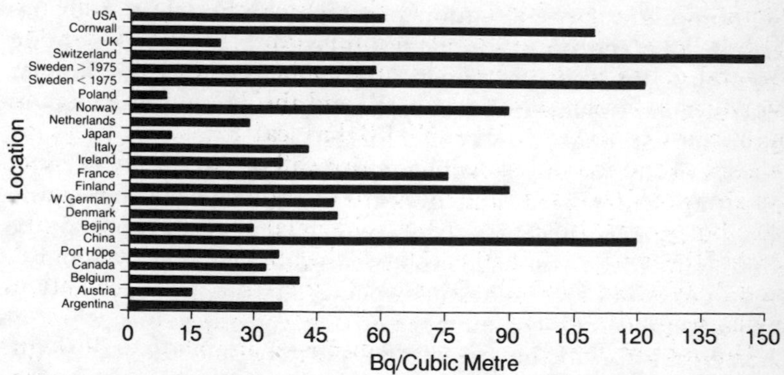

Figure 6.3 Indoor Radon Concentrations Around the World. (Mean or Median Indoor Measurement.)

diated and to what extent. A corollary of this is an assessment of the relative radiosensitivity of the various cell populations in the lung. The point of the exercise is to establish a dosimetric model that will link the concentration of radon in the breathed air, which may be measured, to the dose to the lung and the effect in terms of increased incidences of lung cancers that this dose might produce.

After several years of discussion, most lung models agree within factors of about two to five. All agree that the sensitive cells – those at risk – are the cells lining the bronchi in the mid-part of the lung. It has also been established that after inhalation of radon and its daughters the dose to these cells is about five to ten times higher than the dose to the deep lung, the gas exchange or alveolar region. This is an important finding because it has also been established that this ratio of doses may not be the same for other important inhaled materials, e.g. plutonium. What has not been really decided is the volume of tissue or cells over which the dose must be averaged. Because of the inhomogeneous distribution of dose in the lung after radon daughter inhalation, there are other problems in assessing the effective dose equivalent. The effective dose equivalent is found by multiplying the lung dose equivalent (assuming a quality factor for alpha radiation of twenty) by a risk weighting factor. The risk weighting factor for the lung given by the ICRP (0.12) was set after considerations of whole-body exposure with external radiation. It was in fact derived mainly from data on the excess incidence of lung cancers in Japanese bomb victims at Hiroshima and Nagasaki. This is thought to be inappropriate to the situation in which one part of the lung is irradiated much more than the rest. Because of this, the weighting factor has been split and 0.06 is to apply to each of the bronchiolar and alveolar tissues. In essence, the risk of lung cancer is thought to be proportional to the bronchiolar dose.

Practically all of the epidemiological data on the effects of breathing radon have come from the experiences of uranium and iron-ore miners. Because of this, another important aspect of the dosimetric models derived has been the difference in conditions in uranium mines and domestic dwellings. This applies not only to the breathing rates of miners and the general public when at home, but also to the size of dust particles and the fraction of radon daughters that are attached to the dust by virtue of their electrical charge.

At this point it is worth mentioning a rather odd unit of

exposure to radon that is only gradually slipping out of use. The working level (WL) was introduced in the late 1950s as a convenient measure of the concentration of radon progeny in uranium mine air that could be employed as a measure of exposure. The working level was defined as 'any combination of Po-218, Pb-214, Bi-214 and Po-214 (the short-lived daughters of radon) in 1 litre of air that results in the emission of 1.3×10^5 MeV of alpha particle energy'. This is about the total amount of energy released over a long period of time by the short-lived daughters in equilibrium with 100 pCi (3.7 Bq) of radon (Table 6.3). Only the short-lived daughters are included in this definition because they give most of the dose to the lung. The working level month (WLM) was introduced so that both the duration and level of exposure could be taken into account. Thus the WLM is equal to 170 WL hours, which corresponds to an exposure of 1 WL for 170 hours (1 month). This term is still in use and many epidemiological and dosimetric estimates of risk are based on this unit of potential alpha energy exposure. For use in dwellings, a factor for effective dose equivalent in terms of alpha energy using the models mentioned above is 10 mSv per WLM. It is not, however, practicable to monitor in these units and radon concentrations are now more often given. A consensus from these models is that for general use a conversion factor of 20 Bq per m^3 equivalent to 1 mSv per year could be used.

Giving one figure for this important parameter implies some

Table 6.3 Definition of the 'Working Level'*

Nuclide	*No. of atoms per 3.7 Bq of Rn-222*	*Alpha energy per atom (MeV)*	*Alpha energy per 3.7 Bq (MeV × 10^5)*
Po-218	977	6+7.7=13.7	0.134
Pb-214	8500	7.7	0.660
Bi-214	6310	7.7	0.485
Po-214	0	7.7	0
		Total	1.279

* Explanation of the calculation: 977 atoms of Po-218 are in equilibrium with 3.7 Bq of Rn. Each atom gives 6.0 MeV when it decays plus another 7.7 MeV when its daughter Pb-214 decays. There will also be 8500 atoms of Pb-214: no alpha particle energy will be released until these atoms turn into Bi-214 which releases 7.7 MeV when it decays etc.

degree of certainty which, alas, does not exist. The three most widely accepted lung models (the Harley–Pasternak model, the Jacobi–Eisfeld model and the James–Birchall model) can produce figures that are no closer than about a factor of two to three. The figure given above is in fact near the top of a range given in the BEIR IV report on radon (1988), and is used by the NRPB.

Using the figure given above, what does this imply for the radiation dose to the UK population? (The doses given below are effective dose equivalents, not lung or bronchiolar tissue doses.) The mean radon concentration for in-dwelling locations from the latest NRPB survey was 20.5 Bq per m^3, which implies a mean annual effective dose equivalent in the UK of 1000 μSv (1 mSv). The average person would probably receive another 200 μSv from other indoor locations and 20 μSv from exposure out of doors, giving a total of 1220 μSv per year (1.2 mSv per year when rounded). This compares with a figure of 0.7 mSv per year calculated by the NRPB for their 1984 review. However, annual doses from the radon survey carried out by the NRPB range from 0.4–400 mSv, with the highest mean doses again being in Cornwall (5.7 mSv) and Devon (3.6 mSv). It has also been estimated that doses from radon and its daughters exceed 50 mSv per year (the radiation worker dose limit) in 2000 houses, 20 mSv per year in 20,000 houses, and 10 mSv per year in 50,000 houses. It is thought that in about 6 per cent of the houses in Cornwall the radon dose to the occupants exceeds 50 mSv per year.

In addition to radon (Rn-222), there is also an exposure in the UK to thoron (Rn-220) from thorium-containing rocks and soils. The mean concentration in houses in the UK is about 0.3 Bq per m^3 and this can be converted to an effective dose equivalent of 95 μSv per year. Exposure out of doors contributes another 10 μSv per year so that the mean effective dose equivalent from this source is 105 μSv per year (0.1 mSv, when rounded).

Risk of Lung Cancer

The risk of induction of lung cancer has been investigated in several groups of people exposed generally at high doses and dose rates to low LET radiation (gamma and X-rays) and risk rates have been well established. There are also good data from groups of miners exposed in uranium and other mines to radon and its daughters and some rather equivocal data on the combined effects of smoking and radiation. As is usual with relatively small groups, the epidemiology is limited, but some groups of miners have been followed for many years and complete lifetime experience will be available that has improved confidence in extrapolation models. In fact lung cancer caused by radiation has probably been unknowingly recorded for more than 300 years. A lung disease has been known in miners in the Schneeberg-Jachymov area of Czechoslovakia since the seventeenth century, but was only diagnosed as lung cancer in 1879. Its association with radon was only suggested about sixty years ago when high levels of the gas were first discovered. Early attempts at lung dosimetry thirty years ago identified the likely involvement of radon daughters.

Since that time there have been four large studies of uranium miners and several smaller groups, e.g. fluorspar miners in Newfoundland, and iron ore miners in Sweden, China and the UK. These epidemiological studies now cover about 25,000–30,000 radon-exposed underground workers. The basic data for the four main studies are shown in Table 6.4.

These studies indicate a strong correlation between radon daughter exposure and excess rates of lung cancer. The data from other studies (albeit on a smaller scale) that attempt to link enhanced background levels of radon with lung cancer have been equivocal, but they are being pursued, particularly in Scandinavia.

Estimates of risk factors from these studies are complicated by the need to extrapolate forward in time by either an absolute or relative risk model (see Chapter 3). In addition, an assumption of a linear relationship between radiation exposure to radon daughters and excess rates of lung cancer has been made. However, estimates that have been made by the ICRP and the US Environmental Protection Agency (EPA) indicate lifetime risks of about 2.8 per cent – range 1–6 per cent (ICRP) and 3–13 per cent (EPA), corresponding to a radon daughter level of 20 mSv per year (which is equivalent to a lifetime lung cancer risk rate of 2 in 100 per Sv). This risk is put into perspective in the table from one of the NRPB reports (Table 6.5).

Table 6.4 Basic Data for Epidemiological Studies on Uranium Miners

Quantity		*Colorado*[a] *USA* *1950–77*	*Bohemia*[b] *CSSR* *1948–75*	*Ontario*[c] *Canada* *1955–81*	*France*[d] *1947–83*
Initial number of miners		3,366	2,433	ca.13,400	1,957
Average follow-up period per miner (years)		19	26	15	25.9
Surviving fraction at end of follow-up (%)		72		ca. 80	81
Median age at start of uranium mining (years)		30	35–40	ca. 25	ca. 30
Average working period in uranium mines (years)		9	10	ca. 2	11.4
Number of person-years at risk (PYR)		62,556	ca. 60,000	202,795	50,784
Mean cumulated exposure [WLM]		820	310	60 ± 25	
Fraction of chronic cigarette smokers (%)		ca. 70	ca. 70	50–60	ca. 70
Number of lung	observed	194	ca. 250	82	36
cancer cases	expected	40	ca. 50	57	18.8
during follow-up	excess	154	ca. 200	25	17.2
Relative risk, observed/ expected cases		4.8	ca. 50	1.45	1.9

[a] White miners only.
[b] Study group A only; the total group involved 4,364 miners.
[c] Only uranium miners without prior gold mining.
[d] Exposure data not yet available.

Risk Limitation

The risk of enhanced natural background, particularly from radon daughters, has now been recognized. National and international bodies have suggested standards to be used in future building construction and others to be applied to limit the risk in existing buildings (see Table 6.6). There is clearly some diversity of opinion about the level of emission of radon daughters at which action should be taken. Most countries, however, seem to believe that action is *essential* at about 40 mSv per year (i.e. about 800 Bq

Table 6.5 Approximate Lifetime Risks of Premature Death* from Various Causes Including Lung Cancer from Lifetime Exposure** to Radon Decay Products in Dwellings

Cause	*Risk (%)*
Radon decay products (100 mSv y^{-1})	45
Malignant neoplasms (all causes)	25
Radon decay products (20 mSv y^{-1})	9
Malignant neoplasms, bronchus and lung (all causes)	6
Radon decay products (10 mSv y^{-1})	4.5
Motor vehicle traffic accidents	1
Accidents in the home and residential institutions	1
Motor vehicle traffic accidents involving collision with pedestrian	0.3
Radon decay products (1 mSv y^{-1})	0.45
Conflagration in private dwellings	0.06

Note * Risk calculated using risk rates given in Table 3.9.
** Exposure expressed as effective dose equivalent.

per m^3 of radon, which would be about equivalent to a lifetime risk of radiogenic lung cancer of 4 in 100). It has also not escaped the public's attention that the action level for radiation workers is 15 mSv per year; in the fullness of time this may become the dose limit.

Conclusions

In this chapter, the contributions from natural background to members of the public in the UK have been evaluated and discussed. The range in doses is quite large but for the majority of people the variation from the mean annual doses shown in Table 6.7 is not great.

The exposure of members of the public is dominated by radon daughter emissions. These doses have recently been reassessed in terms of levels and risks, hence the increases shown as averages in Table 6.7. This masks an appreciation of the extreme risks to many people who live in Cornwall and Devon. If biological tissue is unable to distinguish the source of radiation, we may postulate that an increased incidence of lung cancer, at least, should be evident in the southwest. There have been some attempts to identify any health effects in this part of the country, but so far without much success. A suitable control population matched in every way, particularly for socio-economic class and smoking habits, is essential. It must be remembered, however, that the

Table 6.6 Values of Action Levels of Annual Effective Dose Equivalent (mSv per year)

Country or body	*Action level for existing houses*	*Upper bounds for future building*	*Comments*
ICRP	20	10	1984
UK Royal Commission Env Poll	25	5	10th Report 1984
UK NRPB	20	5	1987
Sweden	40	7	Adopted 1980 Target level after action 20 mSv/year
	10	10	Proposed 1984
Norway	10	10	Proposed 1986
Finland	40	10	Adopted 1986
USA	20		Proposed 1984 by NCRP
	8		Adopted 1986 by EPA
Canada	40		Proposed 1985
West Germany	20		Proposed 1986

increased exposure, or collective dose, measured in man-Sv is not that great. The chances of finding a significant increase in a cancer that is already fairly common are therefore not as high as might at first seem likely.

Table 6.7 Summary of Doses to the UK Population from Natural Sources of Radiation

	Average annual dose (μSv)		
Source	*Previous estimate*	*Present estimate*	*Range**
Cosmic radiation	300	250	200–300
External irradiation from terrestrial sources	400	350	100–1,000
Internal irradiation from terrestrial sources	370	300	100–1,000
Exposure to radon and its decay products	700	1,200	300–100,000
Exposure to thoron and its decay products	100	100	50–500
Total	1,870	2,200	1,000–100,000

Note *In some cases, these are little better than order-of-magnitude estimates.

7 Medical Radiation

Is it safe to have an X-ray? This question is rarely asked simply because we assume that the benefits of a diagnostic X-ray outweigh whatever risk is involved. This is one of the very few examples of radiation exposure in which the benefit is immediately obvious. The original question is nevertheless legitimate, especially if X-ray examination is part of a screening programme rather than an aid to diagnosis of disease or trauma. There is thus an immediate problem here in that any serious suggestion of risk might have the effect of making patients wary of radiology. However, it would be difficult for a non-technical person to assess and weigh the benefits and risks of diagnostic X-radiology at the time of the examination.

There are in any case significant ways of reducing the risk to the patient that are easily and cheaply instituted and do not in any way detract from the efficiency and benefit of X-ray diagnosis. These technical and operational modifications are more cost-effective in reducing radiation dose to the general public than some of the civil engineering projects required to do the same thing at nuclear power stations and reprocessing plants. This chapter takes a close look at diagnostic radiology (including screening) and assesses the risks without in any way deriding the benefits.

Discovery of X-rays and Current Techniques

X-rays were discovered by Wilhelm Roentgen in 1895 when he was exploring the passage of electrical discharges through gases at low pressure. Almost by accident, he noticed a glow on a piece of glass covered with zinc sulphide that persisted even when the discharge tube was covered with black paper. His newly discovered rays were obviously not like light and were extremely penetrating.

X-rays are electromagnetic radiation produced whenever high-speed electrons are suddenly slowed down. In a modern X-ray tube the electrons are produced by heating a tungsten wire filament and they are then attracted towards a target by a very large applied voltage (50,000–400,000 volts or more). The target is usually made of tungsten fixed to a copper support, and is angled to allow the X-rays produced to be fired from the tube at 90° from the incident electrons. The tube itself, which is made of glass, is almost evacuated so that gas ionizations are kept to a minimum. Normally the target is cooled and can be rotated to prolong its life.

The energy of the emitted X-rays depends on the accelerating voltage. The X-rays emitted have a more or less continuous spectrum of energies up to a peak (measured in kV) equal to the applied voltage. For most applications it is generally only the higher-energy photons that are useful for diagnosis or therapy. The lower-energy X-rays are therefore filtered out by metal filters placed in the beam. For voltages of 30–120 kV aluminium filters are used; for 100–250 kV, copper; for 200–600kV, tin; and for 600 kV–2 MV, lead.

The principle of diagnostic X-radiography is that certain structures in the body, e.g. bones, which contain high concentrations of elements other than those of which soft tissue is mostly composed (i.e. carbon, hydrogen, oxygen and nitrogen), are opaque to X-rays. As X-rays are to some extent absorbed or attenuated by these structures, the emerging beam can reveal their location, structure and integrity when passed through a patient. The detection of this emerging beam is normally done photographically, although other techniques may be used if continuous surveillance is necessary. An X-ray image, for example, may be seen on a coated screen (fluoroscopy screen) while a barium meal, which is opaque to X-rays, is swallowed.

For maximum efficiency, the beam of X-rays must be collimated ('focused'), and a variety of cones and collimators are available for this purpose. These not only prevent exposure of areas of the patient other than those being examined but also, by minimizing scatter, reduce the dose to the radiographer. Exposure of radiographers is further minimized by carrying out diagnostic radiography in properly designed rooms with shielding and 'maze' entrance systems. Reducing exposure of the patient is slowly becoming a feature of equipment design and is of critical importance.

Equipment Design Features to Reduce Doses

Exposure of the patient is an inherent feature of diagnostic radiology. However small the radiation dose absorbed by a patient, this will increase that person's risk of late-occurring effects such as cancer. It is therefore important to minimize doses while maintaining the efficiency of radiography as a diagnostic aid. If, for example, a reduction in dose meant a reduction of the amount of information that could be obtained from a single X-ray and several repeats were therefore necessary, then the dose reduction would have been nullified. Many of the techniques aimed at the goal of maximum efficiency and minimum dose are universally employed and are a feature of equipment design evolution; others seem to be employed or not at the whim of the radiographer or hospital administration. The following techniques and equipment designs have been or could be used to maximize efficiency and minimize dose:

(a) Radiation quality is a measure of the penetrating power of the X-ray beam and depends on a number of factors. The higher the voltage applied to the tube, the more penetrating will be the emerging X-ray beam. Thus, a better (more intense) image will be received on the X-ray film for a lower skin surface dose to the patient. However, this will be offset by the fact that at higher voltages the contrast between skin and bone is reduced. Thus the tube voltage must be the optimum for the type of X-ray examination to be carried out.

The ratio of beam penetration and dose reduction can be improved by the use of filtration. A filter attenuates the lower-energy component of the X-ray beam that would otherwise be absorbed by the patient. Thus increased filtration results in considerably reduced skin doses. In general, careful control has to be exercised over tube voltage and filtration because as these are increased patient doses are reduced but so is radiographic contrast, particularly for bone. Where bone contrast is not important, however, as in chest radiography and barium contrast studies, high voltage techniques should be used to reduce patient doses.

(b) After the X-rays are produced, the next consideration is the size and alignment of the beam relative to the patient. Here again there are opportunities for dose reduction or, at least, dose control. The smallest possible beam is clearly of benefit to the patient since it minimizes irradiation of sensitive tissues such as the bone marrow and reproductive organs (gonads). Poor techniques in chest radiography can, for example, irradiate the female gonads and even the male gonads. If care is not taken, it is also possible to aim the primary beam at the gonads when, for instance, radiographing the hand. Collimators that automatically restrict the beam to the size of the film cassette in the machine are now used. Older X-ray machines, however, often had circular cone collimators that wasted some of the beam outside the square film.

Shielding of some areas of the body produces a dramatic reduction in dose to tissues. This is particularly true of gonad shields for men and, to a lesser extent for women. Gonadal shielding in males can reduce the dose to the gonads by 95 per cent when the gonads are in the direct beam; in females the reduction is about 50 per cent. Ovarian shields can only be used effectively by experienced radiographers but for most abdominal examinations, such as lumbar, spinal, pelvic and excretory urograms, male gonadal shields are effective. Members of the public need to be encouraged to insist on such rudimentary precautions being used. Some radiographic examinations, e.g. cerebral angiography, produce large doses to the eyes that can be reduced by eye shields. For example, doses reduced by ten times can be achieved by using specially designed lead glasses. These shielding devices and positioning techniques are available and are used by experienced radiographers who have a regard for risk.

(c) Once the primary X-ray beam has passed through the patient, as much use must be made of it as possible in order that the dose to the patient is kept to a minimum. Grids are often used between patient and film to reduce scattering. These will absorb some radiation, so alignment of the beam and the grid is of importance. Again, the age and design of the X-ray equipment will dictate the efficiency with which the beam is delivered to the film. The equipment needs to be well maintained and as up-to-date as possible. Unfortunately hospital budgets will often place limitations on technological updates that could limit the risks to patients.

Costs may restrict the use of slit radiography, another method of reducing patient doses. In this technique, which is not suitable for all examinations, the primary beam is directed through one or more slits that are moved across the area to be radiographed (with a coincident moving slit between patient and film). Contrast is improved and patient doses can be halved. The longer exposure time required makes slit radiography more difficult, especially with children and in casualty cases, but the reductions in dose may well justify the additional cost.

(d) Before an X-ray reaches the film it will necessarily have passed through other materials as well as through the patient. These materials will absorb X-rays, and the use of carbon-fibre composites in radiography offers the possibility of significant reductions of patient doses since they absorb less radiation. Carbon fibres are woven into a fabric and bound with resin, and the solid materials produced have excellent properties. They can be as strong as aluminium for only half the weight and, most importantly, have only about one-eighth of the X-ray absorbance. For example, 3–15 per cent dose reduction can be achieved by the use of carbon fibre in tops for the tables on which patients lie, 6–12 per cent from their use in the fronts for film cassettes, and 20–30 per cent if they are used in grid covers. The only disadvantage to the use of carbon-fibre materials is the extra cost, which may be as much as 50 per cent, although prices are now falling.

(e) For all diagnostic purposes screen-type film is used. The intensifying screen consists of a thin layer of tiny phosphor crystals coated onto a cardboard or plastic support. Each crystal absorbs X-ray energy and emits an amount of light in direct proportion to the energy absorbed. If an intensifying screen is placed between patient and film the image is intensified on the film and a smaller dose can be used for the same result. This technology, originally suggested in 1902, gradually came into use and by the 1940s was common. Intensifying screens have slowly improved, but more recently the introduction of so-called rare-earth screens, rather than the conventional calcium tungstate screens, has led to a dramatic reduction, about 50 per cent, in patient dose. As costs are similar, they should now be in general use. Amazingly, however, this is not so and for some inexplicable reason only about half the radiography departments in the UK report the use of such equipment. This should be part of the optimization (the application of ALARA) of radiation protection implicit in the Ionizing Radiations Regulations (1985) in the UK. It is mystifying why the medical profession is allowed to drag its feet in this way when the nuclear power industry would so rapidly be called to task.

(f) When the primary beam, suitably intensified, finally reaches the film, the lowest patient dose is best achieved by recording the image on fast film. The first X-ray films were glass plates but by 1920 these were becoming a rarity. In 1914 Kodak introduced a large (11 inches × 17 inches) single-coated film that had greater sensitivity than any glass plate. One early problem was the flammability of the cellulose nitrate base but this was eventually superseded by cellulose acetate and then polyester. Development of film technology has meant that patient doses have been reduced by about 75 per cent since the days of glass plates. Much of this improvement, however, can be nullified by bad processing of the film. For instance, a two-degree reduction (2.3°C) in the temperature of the film developer can effectively reduce the film speed by 20 per cent. It has been estimated that processing X-ray films at the correct temperatures could result in such an improvement in image quality that it would theoretically result in a collective dose saving equivalent to 140 times the annual collective dose resulting from the sea discharges from Sellafield. This sort of assessment is hedged with uncertainty but it certainly gives a feel for the simple procedures that could make large positive savings in collective dose (and therefore collective harm) in the UK. The inescapable conclusion is that the risk of cancer has been increased by bad radiographic practice.

(g) There are a number of less obvious ways in which dose savings could be made on a nationwide scale. More intensive and specialized training could, for example, result in an appreciation of risk and the need for control of processing. The number of wasted or rejected films is also of some importance; it has been estimated that at least 5 per cent of exposures are wasted and therefore have to be repeated, with a consequent doubling of patient doses.

Fluoroscopy

A radiographic examination often has to continue while a procedure such as insertion of a catheter is carried out. For such an examination the X-ray image is received on a fluoroscopy screen and is either viewed directly or, preferably, remotely on a TV screen. The length of time that a fluoroscopic procedure takes and thus the resulting dose varies widely, depending on the type of examination, the cooperation of the patient and the skill of the radiographer. In the UK, for example, barium meal examinations were recorded as having lasted between 1 and 620 seconds (a mean of 146 seconds) in one survey. In general, fluoroscopy examinations produce larger patient doses than other diagnostic tests and, in principle at least, there is therefore a greater opportunity to reduce doses. There is certainly an optimum time for a particular technique and this can only be achieved by skill and experience. This is particularly true of longer techniques, which can give appreciable doses to radiographers, theatre nurses and patients. Fluoroscopy times for coronary angiograms, for example, are typically 10–20 minutes and the skin dose can be 0.5–1 Sv! Fluoroscopy seems to be used widely in developing countries where doses of a few tens of mSv are common for chest and upper gastro-intestinal tract examinations. However, it is possible to combine radiography with fluoroscopy and achieve a dose reduction. Nevertheless, screening times do not seem to have reduced over the last few years and the introduction of automatic brightness controls to allow screen viewing in ambient light has increased patient doses. There are reports of an appreciable collective dose attributed to fluoroscopy when this is used merely to position patients prior to radiography.

Computerized Tomography

In computerized tomography (CT) scanning, a narrow (1–2 cm wide) fan-shaped beam of X-rays is used. The X-ray tube rotates around the axis of the body during the investigation so that a 'slice' of the body receives the direct beam. A series of electronic radiation detectors, rather than a film, is used to measure the transmission of X-rays through the patient and this information is gathered and processed by computer. The CT scanner is a powerful tool in skilled hands. The exposure times and skin doses can be higher than those received in conventional X-radiography but the information received is much more than from a single X-ray. Of the collective dose of 17,000 man-Sv per year received in the UK from medical radiation, 500 man-Sv are from CT.

Dental Radiography

Dental radiography is the most common type of diagnostic X-ray examination and its use is increasing rapidly; it more than doubled between 1970 and 1983. In 1983 there were 156 examinations per 1000 of the population, a total of about 9-million examinations, and about 18 per cent of these consisted of pantomography (all-mouth examinations). However, the effective dose equivalent contribution from dental radiography is only about 200 man-Sv, which is only about 1.25 per cent of the total from diagnostic radiology. Apart from ensuring that only essential X-ray examinations are carried out, there seems to be no easy or useful way of reducing this dose contribution.

Nuclear Medicine

The use of radioactively labelled pharmaceuticals for diagnostic purposes is also on the increase. In 1982, an NRPB survey showed that 380,000 procedures took place in the UK, 62 per cent of them using Tc-99m. The overall collective dose to patients was 950 man-Sv. Since 1983, the Tc-99m use has increased such that 75 per cent of all examinations of this type use it and the collective dose is now about 1,000 man-Sv. This corresponds to an average annual per caput dose of 20 μSv (about 6 per cent of the total). However, this average obscures some high individual doses, particularly for imaging techniques. Nevertheless, apart from a few effective dose equivalents up to about 40 mSv per examination, the majority are less than 10 mSv and for non-imaging they are generally less than 0.5 mSv. The highest organ doses (as compared to effective whole-body doses) may be as high as several hundred mSv, but they are mostly below 25 mSv. Generally, attempts are made to reduce the doses to patients and staff by choosing radionuclides that emit lower doses – for instance, the use of I-125 or Tc-99m rather than I-131 for thyroid imaging – and the use of extremely sensitive detection equipment.

Screening Procedures

Screening surveys are currently being carried out in many parts of the world. The two most important are screening for breast cancer (mammography) and screening for lung cancer.

Mammography

Mammography is a highly specialized radiographic technique that involves considerable emotional factors. There are three ways in which breast cancer may be detected, only one of which is really invasive. First, physical examination can detect a lump or thickening of the breast tissue by palpating or feeling with the fingers. This does no harm but is not a very reliable or sensitive method. Secondly, thermography can be used to record the pattern of heat given off by the breast, the point being that the temperature of the skin over a tumour is hotter than surrounding tissue. Like physical examination, this technique is harmless and has potential but it is as yet unproven. The third technique is mammography, in which low-energy X-rays are used to produce the usual shadow picture of tissues. Because low-energy X-rays are used, abnormal structures or nodules (concentrations) of cells show up. There is no doubt that this technique is effective and its use will detect cancers earlier than the other procedures. Against this should be set the fact that an appreciable radiation dose is given to the breasts and there will be an increased risk of cancer in all those irradiated. If millions of women are screened in this way, a far from negligible incidence of breast cancer could result. Thus, screening of this type should only be done under ideal conditions by the most experienced radiographers so that risks are minimized. In simple terms, mammography as a routine screening technique should only be applied to women over 35 years of age. Younger women should only be examined if there is some definite reason to do so, for example, the presence of a lump. The incidence of breast cancer is very much lower in women under 35–40 and so the imaging of breast cancer by this technique is much better in the post-menopausal breast than in the premenopausal one. This is not a test easily carried out by an average GP since it requires personnel with special expertise and experience.

The actual dose involved in a mammographic test is variable. A series of films could involve a dose of about 10 mSv (1 rem) per film. A typical examination requires two films for each breast and therefore doses of 20 mSv (2 rem) to both breasts are likely. However, doses vary a good deal. In Italy, for instance, they range from 0.18–11.1 mSv per film, in Germany from 0.8–35 mSv, in the Netherlands from 1.0–8.8 mSv and in the USA from 2.5–11.6 mSv. In the UK the mean dose is about 1.7 mSv.

The risk of breast cancer at 20 mSv is 1 in 10,000, which converts into 100 cases in 1-million women. In the mid-1970s there was a panic among women in the USA over the incidence of breast cancer which was perceived to be increasing. Several million women demanded mammography and many under-trained physicians carried out examinations of young women. Often the films produced were unusable. This widespread use of ineffective radiology was condemned as totally irresponsible and has been described as 'a clear illustration of the uncontrolled use of radiation which is more dangerous to the public than all the testing of nuclear weapons and the generation of nuclear power'.

Other screening tests

The other screening tests carried out on a large scale (leaving aside pre-employment radiography, which is very variably carried out) represent a much smaller collective dose and risk than mammography. This was not always so; mass chest radiographic screening has largely been abandoned in the UK because the benefit (such as the early detection of lung disorders including cancer) was outweighed by the risks. The procedure persists elsewhere, however, and doses are quite high, e.g. 0.8–1.5 mSv per examination in Japan and up to 0.98 mSv in France. The Japanese also have mass screenings for stomach cancer; 4 million such examinations were carried out in 1980, the collective dose being 16,000 man-Sv. Certainly the whole subject of mass screening is controversial. The issue is whether early detection of, say, a lung cancer improves a patient's chances of survival. It certainly does not justify exposure of people showing negative results to the risk of radiation-induced cancer. Perhaps such routine X-rays should be restricted to heavy smokers or those in high-risk employment. In the USA, the EPA has long recommended that chest X-rays should not be part of pre-employment medical tests. However, the pressure to perform X-rays for medico-legal reasons, i.e. to avoid claims for lack of diagnostic care, are strong and in private medical practice they also represent a substantial source of revenue. The risk–benefit balance is tipped away from the individual in such cases.

Radiology of Premature Babies

Premature babies sometimes stay in neonatal intensive care units for periods extending for weeks and sometimes months. During this critical part of their lives, they often undergo numerous X-ray examinations. Some neonates receive 100–250 X-ray examinations, mostly of the chest, which, because of the small size and inefficiency of available shields, may well include a greater part of the torso. They may also receive barium examinations and CT scans. The bone marrow dose has been found to vary approximately inversely with birth weight. Often, it has been found that 25 per cent of the exposure from such examinations was contributed by test exposures to perfect the technique!

There have been a number of surveys to ascertain the current position, one of the more recent (1986) having reported experiences at a hospital in which 350 premature babies had been treated in one year. These babies received an average 6.5 radiographs but the range was very large – 155 received only 1 X-ray, but 18 had over 25 (the mean of this group of 18 being 59 X-rays). In addition, examinations involving fairly large doses included six fluoroscopies, six cardiac catheterizations and two CT scans of the brain. The doses involved in these examinations have also been estimated (see Table 7.1). However, these doses may well be lower bounds because they have been estimated under ideal conditions in a large, well-equipped hospital. Radiographs in the neonatal period are usually to assist in the diagnosis of respiratory difficulties, i.e. to check for pneumothorax or a collapsed segment of lung. These and other radiographs to check the position of catheters are clearly important but must be balanced against the risk of inducing childhood cancer or even genetic defects. In the study reported above, the risk of developing neoplasia in the

Table 7.1 Radiation Dose Exposure During Different Procedures (Premature Babies)

Examination	*Skin dose/Exposure*
Premature babies (chest + abdomen)	0.07 mGy
Skull (babies)	0.12 mGy
Fluoroscopy	1.3 mGy/min
Cardiac catheterization	131 mGy (mean) Range 52–210 mGy
Computerized tomography	62 mGy (mean) (7.75 mGy per slice)

group of highest-exposed babies (average number of radiographs, 59) can be estimated to be about 1 in 4000. These babies had sufficient respiratory distress to require ventilation, but in such cases the mortality is about 19 per cent. Thus, there is no question that radiology presents a lower risk and undeniable benefit as long as it results in some saving of life. Although the arguments presented give no grounds for unnecessary radiographs of the newborn, they also suggest that the induction of cancer is not a really strong argument for avoiding such radiography. For cardiac catheterizations and CT scans, the doses (and risks as high as 1 in 300) are much higher. It seems sensible therefore to make strict selection of patients a criterion for use of CT and give consideration to the use of ultrasonagraphy for cardiac catheterization.

Clearly radiography of the newborn should be carried out with the correct collimation and shields and with a full awareness of the possible harmful effects. However, the risks of not using it in intensive care outweigh the risks of either cancer or genetic damage.

Doses Received in X-ray Examinations

The annual collective dose from medical radiology in the UK was estimated by the NRPB in 1984 to be 14,000 man-Sv and in 1988

17,000 man-Sv, an increase of about 21 per cent in four years. This is also reflected in the average annual per caput dose, which increased from 250 μSv to 300 μSv.

The latest (1986) NRPB survey of doses received for various simple and complex diagnostic X-ray examinations in a representative sample of hospitals makes alarming reading. The results are presented in Table 7.2 as the mean effective dose equivalents, the effective dose equivalent being the sum of the risk weighted dose equivalent to the various body organs that were irradiated.

There is no doubt that the mean absolute levels of effective dose could be reduced by universal use of the techniques already outlined, but what is really appalling are the ranges of doses for the same examination. As can be seen from Table 7.2, the ranges are often in excess of 100 between hospitals carrying out what are nominally the same examinations. These variations are not easily explained away and surely require the most urgent attention of the government, particularly in regard to optimization of patient doses which is required under the Ionizing Radiation Regulations of 1985 and Directives from the Commission of the European Communities (1985).

However, it should be noted that the age range data for

Table 7.2 Summary Effective Dose Equivalents (mSv) for X-ray Examinations of Adults

Examination	*Dose (mean)*	*Dose range*	*Age (mean)*	*Age range*
Lumbar spine	2.15	0.4–7.4	53	16–91
Chest	0.05	0.01–1.3	55	16–93
Skull	0.15	0.01–0.5	51	16–99
Abdomen	1.39	0.12–9.9	58	16–100
Thoracic spine	0.92	0.2–4.4	56	16–88
Pelvis	1.22	0.1–5.8	62	16–99
IVU*	4.36	1.4–35.1	52	17–94
Barium meal	3.83	0.6–24.4	56	16–97
Barium enema	7.69	2.9–33.6	59	18–91
Cholangiography†	2.59	0.4–9.1	58	25–83
Cholecystography†	0.59	0.13–5.0	49	18–88
Cholangiography/ cholecystography	1.17	0.15–9.0	50	18–88

Note *Intravenous urography.
†Radiography of the gall bladder and bile ducts

treatment indicate that radiography was in most cases given in late middle or old age (see Table 7.3). For this reason, it is arguable that risks should be age-weighted to be truly representative. The only radiography that could be totally disregarded in terms of risk to the patient is that received in terminal illness. In fact, it has been estimated that this exposure accounts for less than 5 per cent of total radiography, and therefore any correction to risk estimates from this is very small.

Alternatives to Conventional Radiography

The rapid rise in X-radiography suggests a review should always be made of alternative techniques that provide as much or even better information with much less or no risk to the patient.

There has been a sharp increase in digital techniques that

Table 7.3 Incidence of X-ray Examinations in the UK (Patients in Each Age Group as a Percentage of the Total Number of Each Sex Receiving X-rays)

Age range	*Male*	*Female*
0–4	4.0	3.3
5–9	3.3	2.7
10–14	5.8	4.8
15–19	7.9	5.6
20–24	7.8	6.3
25–29	5.8	4.6
30–34	6.3	4.7
35–39	7.3	6.6
40-44	5.6	5.9
45–49	5.3	5.8
50–54	6.6	6.8
55-59	6.9	7.1
60–64	7.7	7.0
65–69	5.9	7.0
70–74	6.7	8.1
75–79	4.7	6.8
80–84	2.3	5.6
>84	1.2	3.9

Source Data by courtesy of the National Radiological Protection Board.

utilize a system of recording transmitted photons on an image intensifier or similar device other than film. This process allows computer manipulation of the images and could result in lower doses than conventional X-radiography or fluoroscopy.

Since 1982, nuclear magnetic resonance imaging (NMRI) has also become available. Here, the images are created using radio waves and magnetic pulses; no ionizing radiation is used. At present, this technique is mostly used for brain and spinal cord imaging.

The use of ultrasound as a replacement for X-rays continues to increase very rapidly. The number of hospitals using this technique went up by a factor of 20 between 1973 and 1979, and has doubled again since 1979. The largest proportion of examinations are gall bladder, abdominal and particularly obstetrical. This technique is also being used for neonatal head examinations as a substitute for cranial CT.

Clearly although techniques that could replace X-radiography are becoming available, they are generally expensive and are not in widespread use. Indeed, X-rays are still used in more than 90 per cent of examinations. However, with the risk to the patient in mind, it would seem a matter of paramount importance to use such techniques wherever possible.

Usage of X-rays and Future Trends

The present global availability of X-ray diagnosis is very varied but information on this is almost totally lacking from 50 per cent of the world. One X-ray machine is shared by fewer than 2000 people in one country and by more than 600,000 people in others. The frequency of use is also varied, 15–20 examinations annually per 1000 people in some countries and 1000 per 1000 in others. The latest UNSCEAR report estimates that of the 5-billion or so people in the world, more than three-quarters have no access to X-ray diagnosis. In some countries, 30–70 per cent of X-ray machines are out of order or faulty. For instance, it has been reported that the current and voltage indicators did not work on many machines in Bangladesh and over 40 per cent had no collimator. In India 20 per cent of machines showed excessive leakage of radiation, and had neither cones nor collimators. However, there is little doubt that worldwide use of radiography will increase. One of the reasons for this is the ageing of the population, particularly in Europe. Several European countries expect to have more than 20 per cent of their population over 60 by 2000. These people have a disproportionate number of medical diagnostic and radionuclide investigations. This also applies to the USA where the 31 per cent of the population aged 45 years or older receive 51 per cent of the X-ray examinations and the 11 per cent over 65 years receive 25 per cent of all examinations. However, in Iran 50 per cent of all X-ray examinations are performed on those aged under 30, mainly because only 16 per cent of the population are over 45. Again, in China 44 per cent of chest examinations are performed on those under 30, whereas in the West the comparable figure is 20 per cent. The world's population is experiencing a marked contraction of 0–30-year-olds and an expansion in numbers of those over 30. The oldest populations in 2025 will be in Europe, East Asia, and North America, where X-ray diagnosis is most readily available, and the youngest in Africa and Latin America, with median ages of 22.8 and 27.4 years respectively.

Another reason is simply because as the populations increase, the amount of X-ray diagnosis must increase. World population was 2.5-billion in 1950 and is projected to be 6.1-billion in 2000 and 8.2-billion in 2025. Thus the collective dose could increase by 60 per cent from 1988 to 2025 if no efforts were made to reduce doses.

A further reason for an increase in X-ray use is urbanization. At present 41 per cent of the world's population is classed as urban; by 2025 this will be 65 per cent. Urban populations generally have a much higher frequency of X-ray examinations. UNSCEAR predicts that if 50 per cent of the world's population were urbanized by 2000, if it aged as predicted, and if it totalled 6-billion, the per caput doses and collective doses (and therefore harm) could be 50–100 per cent higher than at present. Thus the risk from diagnostic radiology must be taken seriously. The only countervailing factor is that an *age-weighted* dose equivalent would assume more importance as the irradiated population aged because the risk would have less time to be expressed. Additionally, with the average life expectancy increasing as well, it is difficult to assess how much less effect there would be, because more X-rays are being given to older people.

From the data presented in this chapter it is clear that roughly a quarter of the annual collective dose of 17,000 man-Sv could be saved by the various procedures described. Thus on currently accepted risk rates, the risk of about 250 extra cancers could be saved annually. The saving of 1 man-Sv in this context costs about £40. At a time when some countries, e.g. Italy, are using rare-earth screens for all examinations, the UK is using them in less than half! The UK government is also lukewarm in its insistence that such measures to reduce dose should be introduced. The most recent Guidance Notes for the Protection of Persons from Radiations Arising from Clinical Use (1985) advise that such dose saving equipment 'should be used'. Whether or not the collective dose from the use of diagnostic X-rays increases as expected, there should be an obligation that the ALARA or BPM principle be adopted in radiology in the UK.

This is an example of a situation in which a clear net benefit can be derived from a modest expenditure. The cost of saving a man-Sv in this situation is far less than the cost of saving a man-Sv in the nuclear industry. If the Department of Health find it so difficult to produce the money, maybe the nuclear industry should be asked to volunteer a contribution to save *some* man-Sv cheaply.

Finally, one oft-quoted statement is relevant here: 'Radiation protection is a matter of numbers and some of the numbers do not make sense'.

8 Risks from the Transport of Radioactive Materials

Introduction

Radioactive materials, including waste, have to be transported both nationally and internationally. This attracts considerable media attention and can be a highly visible operation. The necessity of transporting tracer and therapeutic sources for hospital use is clearly understood, but the need for worldwide shipments of nuclear materials needs adequate justification. For most people, their nearest proximity to nuclear fuel or waste may be while it is being transported and fear of an accident is common. As accidents, spillages, etc. seem to be a frequent feature of transport in the modern world it is reasonable to expect a risk of radioactive contamination from them. However, the nuclear industry has identified this risk and gone to extreme lengths to reassure the public on the robustness of its transport containers. Whether these reassurances are convincing is examined in this chapter.

The materials transported consist of:

(a) Irradiated materials and isotopes produced for medical, research and industrial purposes;
(b) imported uranium ores;
(c) processed uranium compounds;
(d) fabricated nuclear fuel elements;
(e) spent fuel returning for reprocessing; and
(f) radioactive waste arising from the nuclear fuel cycle and research.

Because of media coverage one might be forgiven for assuming that most transport of radioactive materials was within the nuclear (power) industry. This is not so. Including the movement of site (industrial) radiography sources, category (a) transport of radioactive materials represents about 77 per cent of all journeys and about 73 per cent in terms of distance travelled, in the UK. The amounts of activity carried per journey are of course relatively small compared with some other categories and generally come into the class of 'low-level radioactive materials'. However, the transport of radioactive materials for medical and industrial use (category (a)) gives rise to about 90 per cent of workers' radiation doses from transport under normal conditions.

More than 500,000 packages containing radioactive sources are transported every year, destined for hospitals and industrial firms; the great majority of these are in fact on their way to airports and ports. Only about 18 per cent are for use in the UK. Even for these relatively short journeys, the public and indeed the transport drivers and freight handlers have to be protected and this is affected by fairly stringent transport rules.

Transport Regulations in the UK

In the UK the transport of radioactive materials is mainly regulated by the Department of Transport (DTp) although the Health and Safety Executive (HSE) and the Department of the Environment (DoE) share that responsibility. The DTp makes and enforces national regulations that are based on the international regulations of the International Atomic Energy Agency (IAEA). The NRPB advises the government on the applicability of these regulations. The situation is further complicated, however, by the role of the Advisory Committee on the Safe Transport of Radioactive Materials (ACTRAM). This was set up in 1985 to 'provide independent advice to the Secretary of State for Transport and the Health and Safety Commission on the arrangements for the safe transport of civil radioactive material'. ACTRAM would seem to duplicate the NRPB's role and is characteristic of the profusion of governmental departments with overlapping and occasionally conflicting responsibilities. In general, however, the several government departments seem to use the IAEA Transport rules as a basis and require no more vigorous tests or controls, and these are accepted by all countries.

These regulations cover all aspects of transport of radioactive sources including types and construction of packages and limiting external doses. In order to comprehend the degree of control that is required, it is worth looking at these rules in detail.

These regulations are an amended and updated (1985) version of those published in 1973. They have not been totally complied with as yet even in the UK. Their purpose is stated to be 'to establish standards of safety which provide an acceptable level of control of the radiation hazards to persons, property, and the environment that are associated with the transport of radioactive materials'. For the purpose of transport, radioactive material is defined as 'any material having a specific activity of greater than 70 kBq/kg (2 nCi/g)'.

The regulations are designed to protect both workers and members of the public and this is to be done by segregating transported packages (by virtue of shielding or distance) so that certain dose limits are complied with. Thus, it is stated that:

(a) For transport workers in the determination of segregation distances or dose rates in regularly occupied areas, a dose level of 5 mSv (500 mrem) per year shall be used.

(b) For members of the public, in the determination of segregation distances or dose rates in regularly occupied public areas or in areas where the public has regular access, a dose level of 1 mSv (100 mrem) per year to the critical group shall be used as the limiting value. This value shall be used ... with the objective of providing reasonable assurance that actual doses from transport of radioactive materials will not exceed small fractions of the appropriate dose limits.

It is noteworthy that for workers, but not for the public, the limit is one-tenth of the dose limit. However, it must be borne in mind that under normal operations the external dose from a transported package is the only hazard. Under abnormal (accident) conditions, the contents of the package may be spread in the environment and intake of radioactive material may also be a risk. For this reason the regulations pay some attention to packaging not only in limiting external dose but also in terms of robustness.

There are basically four types of packages:

(a) Excepted or exempted packages, by far the greatest number, are those containing quite low levels of radioactivity. A table of excepted packages limiting activities is supplied by IAEA and examples for solid materials are:

I-131	5×10^{-4} Tbq
Cs-137	5×10^{-4} TBq
H-3	4×10^{-2} Tbq
C-14	2×10^{-3} TBq
P-32	3×10^{-4} TBq
Mo-99	5×10^{-4} TBq

The radiation level at any point on the external surface of an excepted package should not exceed 5 μSv/hour. In addition, if transported by post the total activity in each package shall not exceed one-tenth of those given in the IAEA list quoted above.

(b) Industrial packages are used to transport materials of large volume but low radioactivity such as uranium ores. These packages might also include articles having low levels of surface contamination.

(c) Type A packages are typically those in which the total radioactivity is greater than an excepted package but does not exceed one thousand times that activity.

(d) Type B packages contain higher levels of activity than Type A but are normally individually designed to suit the material contained. These packages include those designed for fissile material, large radiotherapy and industrial radiography sources and, especially, for the rail transport of spent nuclear fuel back for reprocessing at Sellafield. These packages or flasks have further requirements regarding temperature and pressure limits before they can be moved (see below, pp. 185–7).

The rules and regulations do not end at the type of package. Labelling and dose rate from the package are considered, the latter in relation to the amalgamation of packages into freight loads. In this context, the transport index (TI) is important. The TI for any package is determined as numerically equal to the radiation dose at 1 m from the external surface in mrem/hour (or mSv/(hour × 100)). The dose rate for uranium and thorium ores is to be taken as:

0.4 mSv/hour (40 mrem/hour) for ores and physical concentrates of uranium and thorium;
0.3 mSv/hour (30 mrem/hour) for chemical concentrates of thorium;
0.02 mSv/hour (2 mrem/hour) for chemical concentrates of uranium other than uranium hexafluoride.

For large loads there are also a number of multiplication factors, but for each consignment there are limits on the total TI, as shown in Table 8.1. However transported, the dose rate on the external surface of a radioactive package must not exceed 2 mSv/

Table 8.1 Transport Index (TI) Limits for Freight Containers and Conveyances

	Limit on total sum of transport indexes in a single freight container or aboard a conveyance			
	Not under exclusive use		*Under exclusive use*	
Type of freight container or conveyance	*Non-fissile material*	*Fissile material*	*Non-fissile material*	*Fissile material*
Freight container – small	50	50	n.a.	n.a.
Freight container – large	50	50	No limit	100
Vehicle	50	50	No limit	100
Aircraft				
Passenger	50	50	n.a.	n.a.
Cargo	200	50	No limit	100
Inland waterway vessel	50	50	No limit	100
Seagoing vessel				
1 Hold, compartment or defined deck area:				
Packages, overpacks, small freight containers	50	50	No limit	100
Large freight containers	200	50	No limit	100
2 Total vessel:				
Packages, etc.	200	200	No limit	200
Large freight containers	No limit	No limit	No limit	No limit
3 Special use vessel	n.a.	n.a.	No limit	As approved

hour or 10 mSv/hour if the conveyance (truck, ship or rail wagon) is for the exclusive transport of radioactive materials. Last in this litany of rules and regulations, the package must be adequately labelled. Examples of labels are shown in Figure 8.1. The TI is used to distinguish packages for transport as shown in Table 8.2.

Special conditions are attached to regulations governing certain modes of transport, for example packages by air must be capable of withstanding temperature ranges of −40°C to +50°C. For normal conditions of transport, Type A packages must withstand four tests: a water spray test, a free drop test, a stacking test, and a penetration test. These merely simulate rough handling during transit but clearly accidental conditions such as fire or severe crushing could result in leakage from an excepted or Type A package. This must be prevented if at all possible in the case of Type B packages. Leakages from these containers will be rather more than a local inconvenience and could result in widespread contamination and possible health hazards.

Table 8.2 Categories of Packages According to Transport Index

Conditions		
Transport index	*Maximum radiation level at any point on external surface*	*Category*
0	Not more than 0.005 mSv/h (0.5 mrem/h)	I-WHITE
More than 0 but not more than 1	More than 0.005 mSv/h (0.5 mrem/h) but not more than 0.5 mSv/h (50 mrem/h)	II-YELLOW
More than 1 but not more than 10	More than 0.5 mSv/h (50 mrem/h) but not more than 2 mSv/h (200 mrem/h)	III-YELLOW
More than 10	More than 2 mSv/h (200 mrem/h) but not more than 10 mSv/h (1000 mrem/h)	III-YELLOW and also under exclusive use

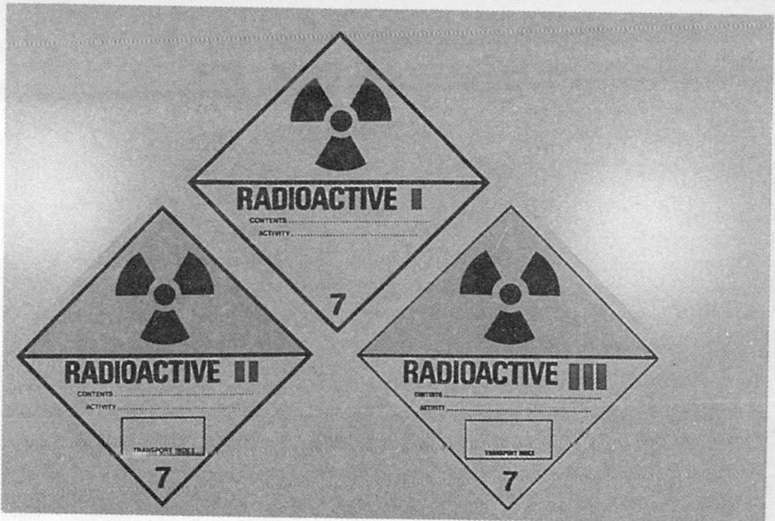

Figure 8.1 Examples of Labels for Radioactive Packages.

Tests for Type B Containers

High-level waste and radiotherapy or radiography sources have to be transported in shielded containers of massive construction (Type B containers, see Figure 8.2) not only to protect people during transit but also to retain their integrity in case of accident. Tests of the construction of these containers are quite stringent as they represent, at least potentially, considerable hazard.

The tests carried out on Type B containers are thought to be necessary in order to ensure that in the event of an accident, e.g. a collision and/or fire during transport, the integrity of the containment will not be breached. Clearly every possible accident scenario cannot be simulated but the tests are rigorous and cumulative.

(a) *Mechanical test* This test consists of three different drop tests. The first is a drop of the container through a distance of 9 m on to an unyielding target. The second is a drop of 1 m on to a vertically mounted bar 15 cm in diameter. The third consists of a 500 kg steel plate being dropped on to the container from a height of 9 m.

(b) *Thermal test* For this test, the container is subjected to a

Figure 8.2 A High-level Waste Container.

hydrocarbon fuel/air fire with a flame temperature of at least 800°C for 30 minutes.

(c) *Water immersion test* For this test, the container is immersed to a depth of 15 m in water for a period of not less than 8 hours. If the package is to contain irradiated fuel, this test is extended to 200 m of water for not less than 1 hour.

Type B containers or flasks used in the UK are subjected to these tests, although one-hundredth scale models are used for the drop tests. It must be said that these rail flasks are robust enough to retain their integrity in most, if not all, accidents it is reasonable to envisage. For other forms of transport, this may not be so.

The Transport of Plutonium by Air

In the UK, plutonium for civil and military uses has been produced for more than thirty years. Some of this material comes from the reprocessing of foreign fuel and has to be returned to its country of origin. In the early 1990s, BNFL will begin to operate its new thermal oxide reprocessing plant (THORP). Much of the finance for this plant has come from reprocessing contracts concluded with a number of European countries (the latest being Germany and Japan). It is proposed that the separated plutonium will be returned to the customer by air. Quite apart from the problem and morality of disposing of, or not returning, the waste from the reprocessing, this way of transporting plutonium seems very risky. However, small amounts of civil plutonium have been transported by air in the UK over the past fifteen years, mostly between Sellafield and Dounreay. Although there is therefore a precedent, agreement for the international air freighting of plutonium has proved to be more difficult to achieve. ACTRAM has reviewed this subject and produced a report. Three considerations led this committee to consider this subject in detail. These were:

(a) The 1986 suggestion from the House of Commons Environment Committee that the transport of most radioactive materials by air should be stopped.
(b) The prediction that the transport of plutonium out of the UK would increase because of contracts secured by BNFL.

(c) Differences in regulations concerning flights by certain countries (notably the USA) in relation to those of the IAEA.

In July 1987 the Trades Unions Congress also expressed concern over this method of transport. It is worth noting, however, that any regulatory changes springing from these concerns about safety would not be applied to military plutonium.

The reasons given for the need for air transport of plutonium relate to security and to a certain extent cost. It is estimated that a sea journey to Japan, for example, could take over twenty days, during which time there would need to be costly security escort. The air journey, by contrast, only takes seventeen hours, and there is therefore presumed to be a saving in security costs. It has also been pointed out that the consequences of an aircraft crash could be minimized by making use of a trans-polar route with more than 99 per cent of the flight over sea or ice. Nevertheless, the USA has now moved from a supportive position to a more pragmatic one, mainly because of differences between its own transport standards and those of the IAEA. The US authorities seem uncertain whether the IAEA test requirements for Type B packages described above would be adequate. Their anxiety may be justified because where plutonium is concerned the integrity of the package must be absolute; there must be no risk of a criticality accident and the leak rate must not exceed a few milligrams a week. It is difficult to see how a package surviving a plane crash without a leak of this magnitude could be devised, and it would be equally difficult to test such a package. The US Nuclear Regulatory Commission (NRC) had stressed that the IAEA drop and fire tests were quite rigorous if carried out properly. In 1975, however, the US Congress passed Public Law 94–79 which imposed a ban on the air transport of plutonium until the NRC had developed a container that would not rupture under crash and blast testing equivalent to the crash and explosion of a high-flying aircraft. Eventually, in 1978, the NRC published an amended set of test criteria for air freight packages. These were much more severe than the IAEA tests. The US fire test, for example, is twice as long as the IAEA's and the impact test about a hundred times more severe. Since the introduction of these tests (published as NUREG-0360) only two package designs have been developed that pass them (PAT-1 and PAT-2). They carry 3.15 kg and 15 g of plutonium respectively. By comparison, the UK package that meets the IAEA requirements (the 1680) holds 50 kg. Japan has since advised the USA that Japanese regulations for air transport will be at least as rigorous as US ones. The fact that packages can, and have been made to conform to these rigorous criteria indicates the seriousness with which air transport of this toxic material is contemplated. Although there are many in the UK who consider these requirements too severe, there are also some who advocate even tighter controls. The consequences of a plane containing plutonium crashing have been considered soberly by ACTRAM, which has assessed the number of additional cancer deaths resulting from the release of 50 g of plutonium – 10,000 times more than the limit imposed by the IAEA – in an urban environment. Using computer models to simulate predicted atmospheric dispersion from a ground release, and a fatal cancer risk factor of 3 in 100 per sievert of exposure, ACTRAM concluded that the additional risk to a person 100 m from the release would only be 1 in 1000. There are those who disagree with the validity of the computer models used and argue that a crash might result in widespread contamination that would be difficult, if not impossible, to clean up. The experiences of the Americans at Palomares and Thule, albeit with much more plutonium, bear witness to the difficulty of plutonium decontamination.

The UK Stance on Air Transport

The UK government appears to be firmly wedded to transport of radioactive materials by air, although its plans may be thwarted by opposition from countries that may be overflown, whose governments consider the US regulations more restrictive than those of the IAEA. In 1987, an all-party Environment Committee reported to the UK government on aspects of the impact of the nuclear fuel cycle on the environment, particularly the management of nuclear wastes including transport. Recommendation 40 of this report stated:

> Wherever possible, radioactive waste, especially spent fuel, high level waste and plutonium should be carried by rail in preference to other modes of transport. The carriage by air of all except the very lowest levels of radioactive materials should be prohibited.

As with the majority of recommendations in this report the British government chose not to endorse this suggestion, with the following statement:

> Although the possible safety advantages of rail transport are recognised, the evidence now available does not clearly indicate any compelling reason to prefer rail transport to other modes on the grounds of safety and for the substances specified. Nevertheless, the safety aspects of the transport of radioactive materials by all modes are subject to periodic re-examination by the Department of Transport taking account of the recommendations arising from IAEA's continuous review process and coming from the regulatory authorities.
>
> The prohibition of the air transport of all radioactive materials, other than those of the very lowest activity levels, would not be justified on current evidence. Large quantities of radioactive materials are moved by air. The risk of an accident is nevertheless very small. If one should occur the packaging required, in accordance with IAEA regulations, has a high standard of resistance to damage, and the possibility of the environment being contaminated is very slight indeed. Air is a particularly appropriate and safe means of transport for many types of consignment and its availability should be maintained. Its continued use, in appropriate circumstances, is also in line with the recommendations of the IAEA, that the time that nuclear material remains in transit should be reduced to a minimum. ACTRAM has suggested the need for a review of IAEA package design standards in relation to air transport. This will be taken up through the normal international procedures for the continuing review of the regulatory standards.

The Transport of Nuclear Materials

The transport of radioactive waste, although not of great significance in terms of miles or journeys, represents the greatest potential risk because of the enormous amounts of radioactivity carried. There is little doubt, however, that by reviewing the need for reprocessing nuclear fuel in the UK and stopping imports of irradiated fuel many of these journeys could be eliminated. Yet there is always going to be a need for the transport of low- and intermediate-level waste to properly managed dumps. This is unavoidable as long as we continue to use nuclear-generated energy, and radioisotopes in medicine, research and industry.

Within the nuclear fuel cycle, the transport of radioactive materials will depend on the relative locations of the fuel source, the fuel refining or separation plants, the reactors, the reprocessing plant and the waste repository. In the UK, there is no mineable source of uranium fuel and partly refined ore has to be shipped in. This material (uranium diuranate, known as yellow cake) is taken by road from ports to the BNFL plant at Springfields, in Lancashire, packed in industrial drums in containers. The next stage of the cycle involves the transport of either uranium hexafluoride or uranium oxide. The hexafluoride (or 'hex') is volatile at normal temperatures and is transported by road as a gas in cylinders containing either 3 or 15 tonnes. The oxide is transported in mild-steel drums clad with cadmium (to absorb neutrons) and hard wood. A full load might consist of more than 200 drums containing just over 5 tonnes of low enriched uranium, again carried by road between BNFL fuel production plants in the north of England. The finished fuel elements for insertion into power station reactors are packed in flat steel boxes in containers. Figure 8.3, which shows the location of BNFL fuel production plants and CEGB power stations in the UK, gives some idea of the distances over which unirradiated fuel must be carried. The radiation doses to drivers, loaders, etc. and members of the public are relatively low at this stage of the nuclear fuel cycle although the possibility of widespread chemical contamination of the environment by catastrophic accident must be considered.

For instance, the freighter *Mont Louis* sank in the North Sea about 16 km north of Ostend in August 1984, after colliding about five hours earlier with a car ferry. The 5000-tonne ship had taken on various materials at Le Havre and Dunkirk and was bound for Riga in Latvia. Its load included 350 tonnes of uranium hexafluoride in 30 containers. It consisted of depleted, slightly enriched and also natural uranium that was to be enriched to between 3.4–3.7 per cent (U-235) in the Soviet Union and then returned to France, Belgium and West Germany to be used as fuel for pressurized water reactors (PWR). The *Mont Louis* sank in 14 m of water but, as the vessel had a beam of 19 m, part was exposed. The ship was reported as having a 'nuclear' cargo, and the world's

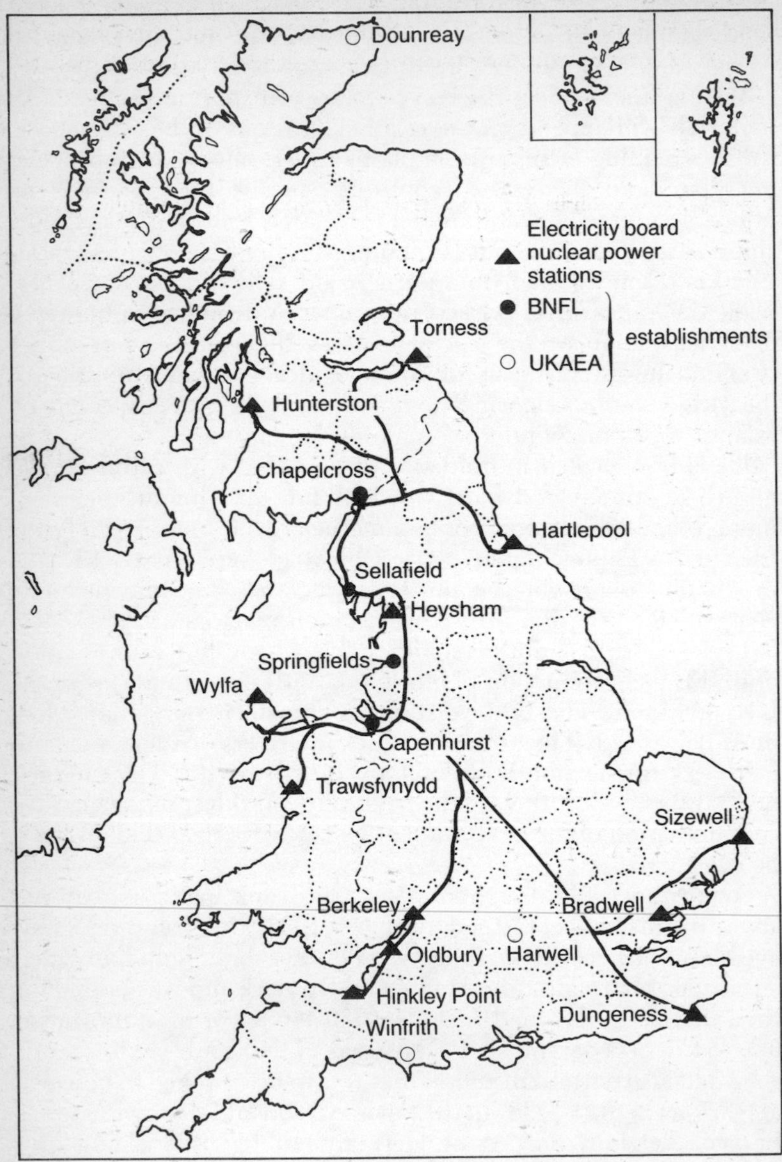

Figure 8.3 Civil Nuclear Sites in the UK.

media took a great deal of interest. A number of estimates of the hazard (both chemical and radiological) were made based on the assumption of leakage from the containers. The work of salvaging the containers, none of which had apparently been damaged, was started a week after the collision. Eventually all 30 of the hex containers and 16 out of 22 empty containers (also part of the cargo) were recovered in about five weeks. Extensive monitoring had been done during and before the recovery operation and no uranium contamination was found. This accident certainly demonstrated the robustness of the hex containers, but it would be inappropriate to draw many other conclusions. There was no fire and none of the containers was damaged in the collision. The recovery was a relatively simple operation in shallow water.

The last part of the fuel cycle, however, has much potential significance from the point of view of hazard. In the UK, irradiated or spent nuclear fuel is transported by rail from the power stations to Sellafield in Cumbria for reprocessing (Figure 8.4). As has been described, the packages (type B) for this transport are immense (up to 100 tonnes in weight) because of shielding and cooling requirements. Although they are also immensely robust, the potential for radiation disaster is ever present.

High level waste from UK reactors is currently stored at the reprocessing plant at Sellafield but waste from the reprocessing of fuel from other countries (such as Japan and West Germany) will

Figure 8.4 Rail Transportation of Nuclear Fuel in the UK.

Table 8.3 Transport Needs in the Nuclear Fuel Cycle for Generation of 1 GW (PWR)

Material	*Amount (Tonnes)*	*From*	*To*
Uranium ore	60,000	Mine	Mill
Uranium yellow cake	170	Mill	Enrichment/fuel fabrication
Fuel elements	37	Fuel fabrication	Reactors
Spent fuel	37	Reactors	Storage/reprocessing
Recovered fuel	25	Reprocessing	Fuel fabrication
High level waste	10	Reprocessing	Waste repository
Other solid wastes	1,000	All facilities	Disposal sites

be returned, so there will be increasing road and rail traffic between Sellafield and BNFL's sea terminal at Barrow.

The total transport needs in the nuclear fuel cycle for generation of 1 GW of electrical energy by a PWR (the new generation of reactors to be constructed in the UK) is shown in Table 8.3. This table, which was compiled by the International Fuel Cycle Evaluation, assumes some recycling of the fuel.

Table 8.4 Averages of the Maximum Dose Rates Recorded in the Vicinity of Irradiated Fuel Transport Flasks

	Distances at which dose rates (μSv h^{-1}) were measured						
Type of flask	Flask sidewall (measured at 1.5m height)					Flask lid	
	Surface	0.5m	1m	2m	6m	Surface	1m
Magnox	14	5.1	1.8	0.64	0.26	20	6.5
AGR	13	9.0	4.0	1.5	0.45	83	27
	Side of wagon (measured at 1.5m height)					Flask surface	
PWR/BWR[1]	n.a.	28	20	11	1.7	83	21

Note [1] These flasks were transported at a higher level above ground than the Magnox and AGR flasks because of the type of wagon used. Measurements were taken at distances from the side of the wagon and from the flask surface.

Types of transported waste

Transported waste is usually solid, simply because that is what is generated. Liquids are far more difficult to contain and transport. 'Dilute and disperse' is therefore unfortunately the general rule, with liquid wastes being drained into the sea.

Solid waste is usually divided into three categories depending on its specific activity (concentration).

(a) High-level waste (HLW) is waste arising from nuclear reactor fuel either as unreprocessed fuel elements or separated fission product waste following reprocessing (at, for instance, Sellafield). This material is not only intensely radioactive but also generates enough heat to melt itself and its containment. Before transportation, discharged fuel elements from reactors are therefore cooled in water in ponds at the power stations for periods of six months or so. Even so, the fuel rods still have to be cooled during transport and require massive shielding. An additional problem with spent fuel is the ever present likelihood of 'criticality', that is, the possibility of the fuel forming a configuration that would sustain a short-lived chain reaction. This might not be of sufficient force to blow apart the spent fuel's containment but it could result in leakage and also a burst of neutrons. This is a major problem to be taken into account when transporting fissile material, particularly as it is now shipped regularly around the world as well as being rail-freighted across countries.

(b) Intermediate-level waste (ILW) tends to be generated at nuclear reprocessing plants and consists of ion exchange resins, nuclear fuel-can debris, plutonium-contaminated material, etc. This waste is not usually heat-producing and is of a much lower specific activity than HLW. However, it cannot be discharged to the environment, so transport to engineered dumps is necessary. There is thus a very strong case for such dumps to be near the sites where waste is generated. This is the category of waste that was dumped, sealed in concrete in steel drums, in the North Atlantic prior to the ban on sea dumping in 1983 (see Chapter 9). This type of waste is also mainly transported in steel drums, sealed in concrete, presumably to limit dispersion in the event of an accident.

(c) Low-level waste (LLW) is usually considered to have a level of radiation sufficiently low to allow it to be judiciously mixed with or diluted into the environment, or dumped in trenches such as the Drigg site in Cumbria. This waste is transported to waste dumps in drums or skips and typically consists of contaminated equipment, clothing, rubble, contaminated soil, etc. This material needs care in transport but any individual load is not likely to present a significant hazard even in an accident.

The transport routes

Routes for waste transport depend entirely on the location of waste dumps (or repositories, as they are often euphemistically termed), and national policies over waste treatment and disposal. The UK has a policy of reprocessing nuclear fuel with storage at Sellafield, an undefined policy for storage of ILW at various sites, and LLW dumps at Drigg (near Sellafield) and at Dounreay. Thus rail transport of HLW tends to be on routes between the nuclear power stations and Sellafield (see Fig. 8.3). ILW and LLW transport routes are more diverse. Japan has no repositories or stores at present (although one is being built at Rokkasho Mura) and intends to have most of its nuclear fuel shipped to Sellafield for reprocessing. Several other countries have contracts with BNFL at Sellafield for reprocessing, so transport of HLW by sea and by road or rail from the port at Barrow (Cumbria) is already common. The West Germans have recently decided not to proceed with a reprocessing plant at Wackersdorf and will rely for reprocessing on a small plant at Karlsruhe, Sellafield and the French plant at Cap de la Hague. The Belgians also use French and British plants. The French have two reprocessing plants, at Cap de la Hague (in the northwest) and Marcoule (on the Rhône) and nuclear power plants spread out throughout the country. They have an LLW dump at the Centre de la Manche near Cap de la Hague. The West Germans use at least two salt mines for LLW and ILW disposal (Asse and Konrad) and the Gorleben site for storage of all types of waste. Thus it can be seen that Europe is crisscrossed with transport routes for HLW and the transport of ILW and LLW must now be commonplace everywhere. The same situation obtains in the USA.

The sea transport of HLW backwards and forwards between

Japan and Europe is now well established, with five ships involved. There has also been some investigation of the possibility of flying the reprocessed plutonium back to Japan.

Transport Hardware

Small radioactive packages

As has already been pointed out, most transport of radioactive materials involves very low levels in exempt or excepted packages. These packages are normally carried in small vans with no shielding. Ninety per cent of them weigh no more than 1–2 kg. The remainder of these packages have a TI (see p. 184 above) of 1 and, maybe because of some shielding, weigh a little more (5–20 kg) but are still manually handled. Less than 1 per cent of the packages weigh more than 30 kg but these will need mechanical handling and may have a higher TI. When loaded into vans, high-TI packages are kept away from the driver's cab. These may have surface dose rates of 100–1300 μSv per hour, giving a dose rate in the driver's cab of 10–80 μSv per hour.

In terms of collective dose to drivers, handlers, etc., the exempt packages make only a small contribution, the significant packages being medical and industrial sources (about 20,000 items per year). From an NRPB survey carried out in 1984, these were shown to consist of:

(a) molybdenum/technetium-99 generators (60 per cent);
(b) other medical and research sources (20 per cent); and
(c) individual sources: iridium-192 radiography sources or caesium-137 sources (20 per cent).

As will be seen, transport and handling of Mo/Tc-99 generators represents a significant irradiation dose even under normal conditions, viz., surface dose rates varying from 500–1300 μSv per hour.

Larger radioactive packages

ROAD AND RAIL

Type B packages for the transport of very large radiography

sources and, more frequently and regularly, the transport of high-level waste (spent fuel rods) need to be massive to provide adequate shielding and to withstand severe accidents. They range up to about 120 tonnes in weight, each having a capacity of about 2–3 tonnes of spent fuel rods.

A typical high-level waste (spent Magnox fuel) 'flask' is shown diagrammatically in Figure 8.5. The flask is essentially an enormously strong steel box with walls 37 cm thick. The 9-tonne lid is secured by sixteen high-tensile steel bolts, each 5 cm in diameter, and the joint is made with two large compression seals. The flasks are fitted with cooling fins and painted white.

Flasks for AGR and PWR fuel are different. The fuel is transported in cylindrically shaped water-filled flasks (Figure 8.6). They are fabricated from steel 9 cm thick and the walls are lined with 19 cm of lead. The unladen weight of the flasks is about 70 tonnes and they carry up to 3 tonnes of spent

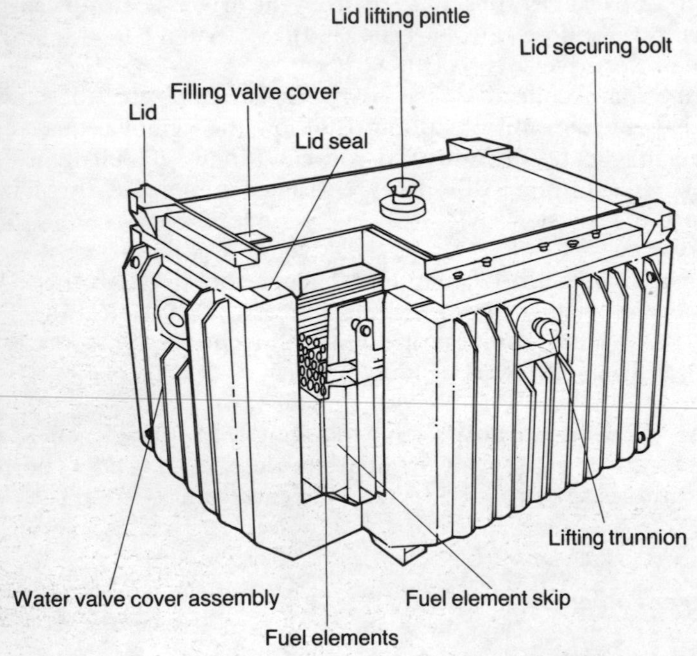

Figure 8.5 A High-level Waste (Spent Magnox Fuel) 'Flask'.

oxide fuel. Typical doses that have been measured by the NRPB on the surface and near examples of these two types of spent-fuel transport flasks are shown in Table 8.4 (see p. 194).

These flasks have to be tested quite rigorously, as has been described in outline earlier (see p. 185 above). The CEGB is conscious that there is public anxiety about the integrity of these flasks and so has carried out two dramatic demonstrations. In 1984, a 48-tonne Magnox fuel transport flask was dropped from a height of 9 m on to an unyielding metal plate anchored to rock in Cheddar, Somerset. The flask fell for 1.5 seconds and reached a speed of about 50 km/h but was reported to be still fully sealed after impact. An even more dramatic test was staged by the CEGB in July 1984 on a British Rail test rail track in Leicestershire. Another Magnox flask on its rail wagon was set across a

Figure 8.6 An AGR/PWR Fuel 'Flask'.

rail track and a 140-tonne diesel locomotive and three coaches were crashed into it at 160 km/h. The locomotive was extensively damaged but the flask remained pressurized although it was pushed 100 m down the track (Figure 8.7). Certainly the demonstrations were good, although albeit expensive PR.

BR uses dedicated trains for the movement of spent fuel flasks. They run at times that as far as possible ensure a direct unhindered journey. The regulations that cover the operation of these trains – which it must be remembered are the means by which members of the public come in closest contact with spent nuclear fuel – are worth recording.

The train must have a guard, who is required to ride in a brake van at the rear of the train. The flask must be separated from the driver and guard by at least 6 m. This may involve the use of barrier wagons of an optimum safe working weight. During the formation of a train, the flask, barrier wagons and guard's van must not be loose shunted, etc. The formation of the train is checked through the Total Operations Processing System (TOPS) – a computerized control system – which checks that staff

Figure 8.7 Results of the CEGB Collision Test of a Fuel 'Flask', July 1984.

have made up the train in accordance with the Regulations, advises error and produces a 'go or no-go' clearance for movement. Train speed is restricted. The driver and guard must have the appropriate written forms of advice as an authority before moving the flasks.

The flasks and wagons must be checked for radioactivity by the consignor, who must produce a written certificate before the start of each journey confirming that there will be no hazard to staff during the course of their normal duties. They are, however, instructed not to linger unnecessarily near the wagons. The BR Scientific Services Unit and staff from the NRPB also carry out random checks on this traffic to ensure compliance with the regulations.

The train's progress must be advised between railway controls throughout its journey and kept under constant notice. Flasks held at stations or marshalling yards must be parked at least 6 m away from places used by people, and wagons containing livestock and dangerous goods; their presence and location must be approved by the BR official controlling the station or yard. After each scheduled stop the guard is required to inspect the train before the journey is resumed. These seem to be the most rigorous controls that can be exercised.

SEA

The carriage of highly active fuel by sea is surprisingly common and, with the signing of contracts between BNFL and Japan and other countries for reprocessing, will become commoner. The first movement of spent fuel between Japan and the UK was in 1969 and PNTL, the shipping company involved, now has five ships specifically designed for this work (Figure 8.8). These ply between Japan, France and the port of Barrow in the UK. Their size is at present limited to 3000 tonnes by the limited access to Japanese reactor ports. They can transport spent fuel flasks, HLW casks or waste drums and have strengthening features, collision bulkheads and additional watertight compartments. Presumably these ships will also carry plutonium recovered by reprocessing if the air transport option is not used.

Figure 8.8 A Specialized Spent Fuel Carrier.

Doses Received Under Normal Transport Operations

The distribution of doses to workers involved in the transport of radioactive materials is shown in Figure 8.9. These are dominated by those incurred during the transport of medical isotopes for nuclear medicine tests (67 per cent). In fact the transport of molybdenum-99/technetium-99m generator leads to the highest radiation doses. Technetium-99m accounts for about 75 per cent of all radioisotope administrations by nuclear medicine departments in UK hospitals. In 1980, about 16,000 generators were transported in the UK, half of which were for export. The surface dose rates were in the range of 0.1–1.3 mSv per hour and the average annual exposure of the road transport workers (256) engaged in their delivery was estimated to be 2.7 mSv (which compares with, say, the average for all medical workers, which was 0.7 mSv). Hidden in this average is a small group of workers (15) whose individual annual dose approaches 15 mSv. Clearly this is not a good state of affairs, especially as these doses could be reduced by better designs of carrying containers. This dose (15 mSv), it should be remembered, has been recommended by the NRPB as a mean annual dose constraint.

Some of these generators are a particular problem when carried by trains with restricted luggage space. For instance, the short half life of rubidium-81/krypton-81m can mean that fairly large activities have to be transported by rail. The combination of segregation distances required by BR and the reduction in the size of goods compartments on some new rolling stock (Sprinter and HSTs) has severely restricted the use of these trains at peak times for this goods carriage.

The transport of waste or spent nuclear fuel is potentially a source of exposure but in reality under normal conditions does not seem to be so. The NRPB has done an extensive survey of the doses received by workers and members of the public from this source. They found that in general dose levels near irradiated fuel transport flasks were low and did not even approach the BR limit of 2 mSv/hour at the surface or 0.1 mSv per hour at 1 m distance. They found that up to 50 per cent of the measured dose was from neutrons from flasks containing PWR or AGR fuel. They also found that there was some slight contamination of the outside of the flasks, presumably from cooling-pond water that had not been removed during decontamination at the power station. Again, this did not reach the limit of 3.7 Bq/cm^2. They found that the highest exposed workers at railheads received less than 100 μSv per hour whereas the majority of rail staff received doses undetect-

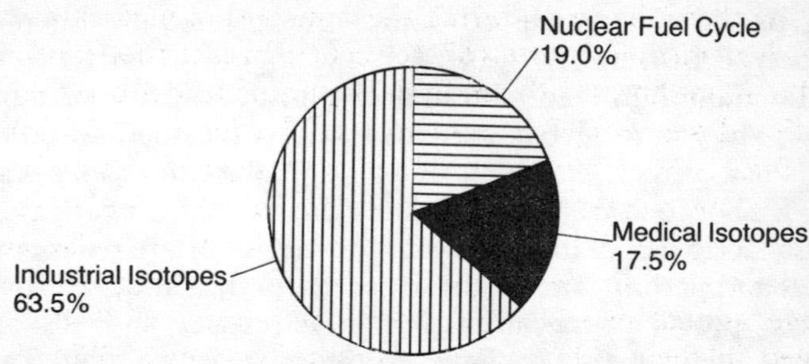

Figure 8.9 Distribution of Worker Exposures as a Result of the Transport of Radioactive Materials.

able from background. The collective dose to rail staff was estimated to be 5×10^{-3} man-Sv per year.

Studies of *public* exposure from the transport of waste and spent fuel have been carried out both by the CEGB and the NRPB. Clearly if the dose to the workers is small the dose to members of the public who live either near marshalling yards or near the routes used for flask transport will be even smaller. In the assessments, several assumptions had to be made about location of the public, occupancy, and airborne release of surface contamination from the flasks. The maximum individual dose to a person living close to a marshalling yard was estimated to be no more than 2 μSv per year. This should be compared with the maximum individual public dose of 40 μSv per year which arises from the movement of medical and industrial isotopes.

Transport Accidents

The transportation of radioactive sources, spent fuel and waste engenders anxiety in the public's mind largely because of the possibility of accidents that might result in spillages and contamination. From what has been said earlier in this chapter, it will seem that this possibility has been faced by the IAEA, CEGB, BNFL and others, particularly in respect of the transport of high-activity materials and plutonium. Assurances are given about the integrity of the transport flasks under quite severe conditions and they do seem to be robust under test. However, any material of this nature must be considered vulnerable when it joins general global transport. Complex emergency arrangements have been drawn up by the CEGB and local authorities to deal with the maximum credible rail accident, the rupture of a spent fuel container in an urban area. Similar arrangements would be put into operation should a plane containing a plutonium package crash. Elaborate analyses have been done of the statistical likelihood of accidents of this nature occurring. They are reassuringly small and perhaps these official estimates should be mentioned with the proviso that most serious industrial and transport accidents that happen have similar low probabilities of occurrence.

The likelihood of impact and severe fire affecting the transportation of PWR fuel from Sizewell to Sellafield has been estimated at about 1 in 1000 million per year. If this accident were to happen in London the NRPB has estimated that the probability of fatal cancer to the closest individual would be about 1 in 4000. This is about one-thousandth of the lifetime natural incidence of cancer. The most severe hazard to a ship is collision and fire. The frequency of this scenario has been estimated for the PNTL/BNFL fleet as about 3 in 100,000 per year.

The industry points out that there is considerable difference between risk as perceived by the public and risk as estimated by calculation (the so called objective risk). Unfortunately, the public is aware that the unthinkable does occasionally happen and the additional complication of an unknown hazard from radiation is difficult to contemplate.

Conclusions

From what has been set out in this chapter, it is evident that under normal conditions of working the hazard from the transportation of radioactive materials is minimal and controllable. Although some workers transporting medical isotopes receive unacceptably high doses, strenuous efforts have been made to reduce the risk from the transport of spent fuel and high-level waste to low levels and they have been successful. Clearly the risk from accidents on the railways, at sea or in the air should be eliminated rather than reduced, but it is unlikely this will ever be done. The risk from rail transportation, for example, is already so small that there is probably no easy way of reducing it further apart from abandoning the reprocessing of fuel and creating dry stores at each power station. In this respect, the proliferation of transport to and from the UK in the future because of the presence and policies of Sellafield is to be deprecated.

Above all, the safety standards of transportation must be maintained, as the public relies entirely on them. The public may therefore have to have blind faith in official reassurances that all is well. The scientist really has a hard time saying that low-probability, high-consequence events will *never* happen, but a responsible public agency would want to be seen to be protecting the public very conservatively. However, there is no real reason for committing large sums of money in excess of that already being spent in reducing the hazard from rail containers. The transport of plutonium by air, on the other hand, needs adequate justification.

9 Radioactive Discharges to the Environment

Introduction

This chapter examines the impact of radioactive discharges to the environment in terms of radiation doses to members of the public. Attention is given mainly to discharges from the nuclear industry because they are by far the largest and, over the past thirty years, have been the most controversial, leaving an obscene legacy of contamination. The discharges of waste from Sellafield will be given particular consideration because, even though in recent years strenuous and successful efforts have been made to reduce discharge rates, this one site is responsible for most radioactive contamination in Western Europe. Up to 1982, BNFL's Sellafield operations were responsible for nearly 99 per cent of liquid radioactive waste discharged into the sea in the UK; by 1985 the proportion was 84 per cent and by 1986 it had, creditably, dropped to 62 per cent.

The Legal Control of Radioactive Waste Disposal

All UK establishments handling radioactive materials on their premises are subject to the Radioactive Substances Act (1960). This Act controls, by authorization under Section 1, both the possession and use of radioactive materials and also, under Section 6, the disposal of waste in any form, whether solid, liquid or gas. In England the authorizations are issued by the Secretary of State for the Environment and, additionally for discharges from the UKAEA and nuclear licensed sites, the Ministry of Agriculture, Fisheries and Food (MAFF). In Scotland and Wales, the relevant Secretaries of State issue authorizations. The Radioactive Substances Act is based heavily on the philosophy of ICRP in that it requires (a) that the discharges of waste must be justifiable in terms of some net benefit; (b) that steps have been taken to ensure that exposures of members of the public to radiation are as low as reasonably achievable, taking into account economic and social factors; and (c) that exposure of representative members of a critical group of the public does not exceed a prescribed dose limit. The implications of condition (c) are examined in more detail below.

The authorizing authorities (DoE and MAFF) usually set numerical limits on the amounts of radioactivity that may be discharged over specified periods of time and also require operators to use the 'best practicable means' to limit the radioactivity of waste. In addition, controls over the range of radioactive species and the manner of their discharge are included. There is also a requirement to measure or monitor the amount of material discharged, and in some cases to monitor the immediate environment and the local environment.

Thus the potential 'polluter' or operator is not only asked to do his own hazard assessment, he is also required to monitor the impact of his own waste. This is clearly not a satisfactory arrangement and so MAFF also monitors the environment to check CEGB and BNFL results, and makes assessments of dose. As will be seen, these checks have not always agreed with operators' results.

The Radioactive Substances Act has some odd clauses, e.g. requiring secrecy of discharge, but is, at present (1989), undergoing revision. With sufficient inspectors to enforce it this legislation could be quite effective. In 1983, a successful case was brought against BNFL for its failure to use the best practical means to reduce discharges that were within their annual discharge limits, but which seriously contaminated the beaches of West Cumbria.

The 1960 Act does not cover sites controlled by the MoD. We have to rely on the MoD to conduct its own monitoring around the Atomic Weapons Establishment, the naval bases at Rosyth, Devonport, the Clyde and Holy Loch and other MoD sites.

However, MAFF also does checks around these sites, the results of which are published annually.

In the UK, all nuclear power stations (except one, Trawsfynydd) and the majority of nuclear plants discharge low-level waste directly into the sea (Figure 9.1) and most of the MAFF monitoring results apply to the aquatic environment.

Where Do the Wastes Come From or Why Are Wastes Discharged?

All discharges of radioactive material in the UK must be justified but, as all industrial, research and medical processes generate waste that it is uneconomic to store, the only debate generally is over where and how disposal is effected.

The nuclear industry discharges by far the greatest amount (in terms of activity) of radioactive material and most of this is created by the current policy of reprocessing spent fuel. We will therefore concentrate on this particular part of the nuclear fuel cycle in this chapter and investigate its impact on the environment.

First, therefore, at what stage of the nuclear fuel cycle does waste arise? The nuclear fuel cycle as it is managed in the UK is shown in Figure 9.2. It concerns all processes from mining of uranium ore through extraction and treatment of that ore, purification of the uranium, enrichment (if necessary), fabrication of the fuel, fission in the reactor, interim storage of the spent fuel, reprocessing and final disposal or storage of the waste.

Uranium is mined abroad, separated as a mixture of oxides, and shipped to this country as 'yellow cake' (ammonium diuranate). The material remaining from these processes, known as tailings, contains most of the uranium decay series, in particular radium and its daughters (see Chapter 6). The safe disposal of this material presents a problem in some countries and the radiation exposure of the local population can be significant. For instance, tailings have been used as landfill in Colorado and radon levels in water and buildings have become a serious problem. The hazard is now recognized, but tailings are a real source of population exposure in, for instance, the USA and Canada.

In the UK yellow cake is further purified at the fuel fabrication

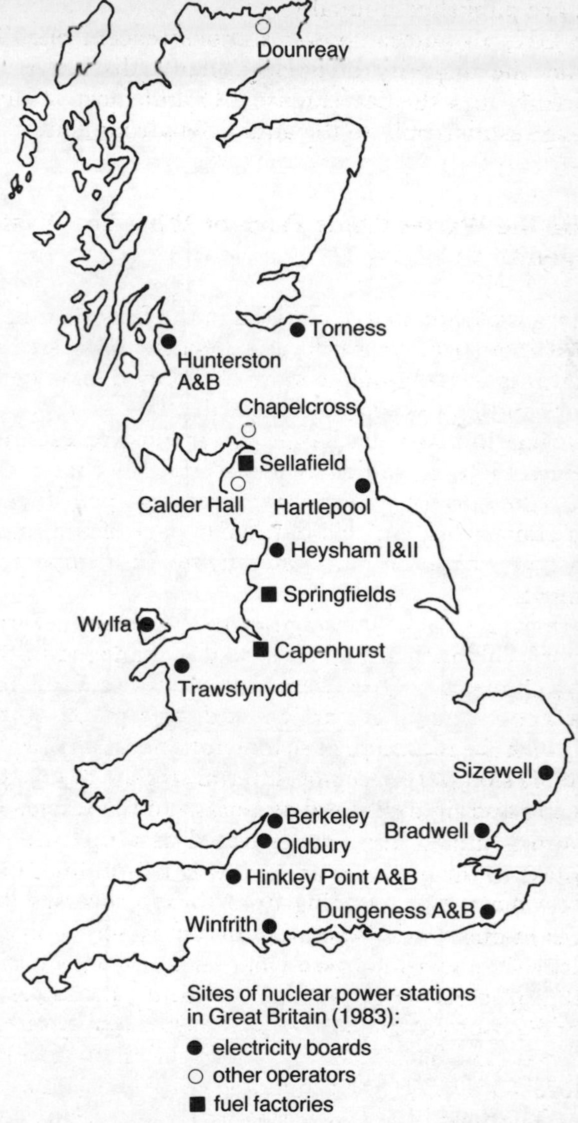

Figure 9.1 UK Nuclear Establishments Giving Rise to Principal Liquid Radioactive Waste Discharges.

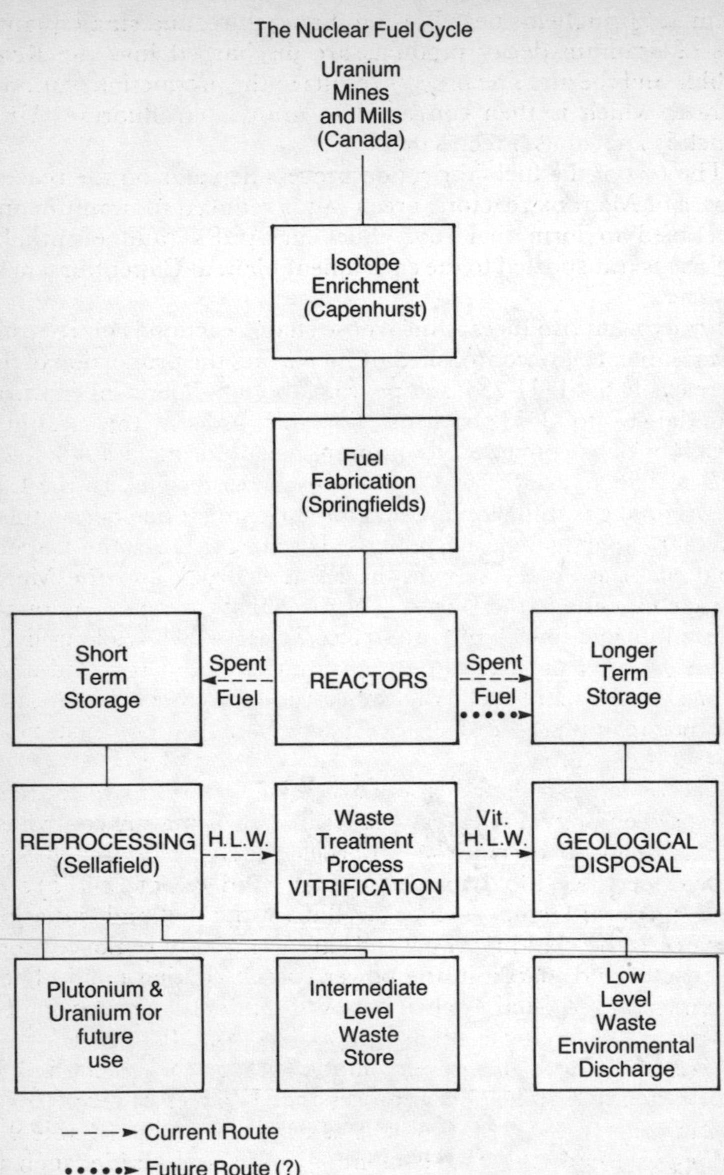

Figure 9.2 The Nuclear Fuel Cycle.

plant at Springfields near Preston, Lancashire, and small quantities of uranium decay products are discharged into the River Ribble and the air. The process results in the production of uranyl nitrate, which is then converted to uranyl tetrafluoride (UF_4) which is known as 'green salt'.

The rest of the fuel fabrication process depends on the reactor type. For Magnox reactors, green salt is reduced to uranium and machined to form fuel rods. Fuel for AGRs (and eventually PWRs) is transported to the enrichment plant at Capenhurst near Chester.

Enrichment produces a fuel whose chain reaction is more easily sustainable. It is accomplished by increasing the proportion of the more easily fissile U-235 isotope from about 0.7 per cent (natural abundance) to 2–3 per cent. For this process the uranium tetrafluoride is converted to uranium hexafluoride (UF_6) known as 'hex' which, above 56°C, is a highly corrosive gas. In the UK the original gas diffusion method for enrichment has been superseded by a centrifugation process. Discharges from the Capenhurst plant go to the sea via the Rivacre Brook and the Meols Sewage Pipeline to the Wirral. The discharges are small amounts of uranium and its daughters and also Tc-99, which is a long-lived fission product present in uranium that has been recycled after recovery from spent fuel. After fabrication into fuel elements and cladding (canning) the fuel rods are sent to nuclear power stations around the country.

Fuel remains in the reactors in the UK for up to two to three years, depending on reactor type. Usable life is mainly controlled by the build-up of neutron absorbers, such as certain fission products, e.g. Xe-135 (which arises from the decay of Te-135) and other, physical factors such as swelling of the fuel and radiation damage to the cladding. Fuel rods are eventually removed from the reactor and placed in the power station cooling pond. They are intensely radioactive when removed but levels are reduced by storage in the cooling pond as can be seen from Table 9.1.

After cooling in the power station ponds, the spent fuel is transported to Sellafield for reprocessing. The fuel still contains a large inventory of fission products as can be seen from Table 9.1 (cooling is usually three to six months), and also actinides such as Pu-239, Am-241 and Cm-242. The purpose of reprocessing (the need for which is discussed later) is stated by BNFL to be to recover unused uranium and separate plutonium, both of which

Table 9.1 Theoretical Amounts of Some Fission Products in the Fuel (in TBq) at Different Times After Removal from a Reactor Operated at 1 MW(e) for One Year

Radionuclide	*Immediately after removal*	*Three months later*	*Five years later*
^{85}Kr	7.1	6.9	4.9
^{89}Sr	1400	380	
^{90}Sr	53	52	44
^{90}Y*	53	52	44
^{91}Y	1800	540	
^{95}Zr	1800	540	
^{95}Nb*	1800	1100	
^{103}Ru	1100	220	
^{103m}Rh*	1100	220	
^{106}Ru	81	67	2.6
^{106}Rh*	81	67	2.6
^{131}I	930	0.15	
^{133}Xe	2000		
^{137}Cs	40	40	36
^{137}mBa*	38	38	34
^{140}Ba	1900	8.5	
^{140}La*	1900	10	
^{141}Ce	1800	180	
^{143}Pr	1700	11	
^{144}Ce	1000	770	10
^{144}Pr*	1000	770	10
^{147}Nd	810	1.5	
^{147}Pm*	180	180	50

Note *Daughters

are used for recycling; up to 35 per cent of refuelling can come from this source. In addition, Magnox fuel that has been stored in water for some time must be reprocessed because of corrosion of the fuel cans. The two reprocessing plants in the UK – the main one at Sellafield and a much smaller facility at Dounreay in Scotland – have processed more than 30,000 tonnes of Magnox fuel and about 100 tonnes of oxide (AGR/PWR) fuel. At present oxide fuel is being stored awaiting the completion of the thermal oxide fuel facility (THORP).

Spent fuel arriving at Sellafield in transport flasks is transferred to skips that are stored for a further period of months to allow for extra decay of fission products. They are then removed by remote handling to a shielded facility and the cladding is stripped away. The fuel rods are dissolved in nitric acid, and uranium and plutonium are separated from fission products by a complex solvent extraction process using tributyl phosphate (TBP) dissolved in odourless kerosene (OK). Early in this process gaseous wastes such as Kr-85 are released to the atmosphere.

Following the dissolution and separation, the nitric acid phase is removed for storage as high-level waste (HLW). This waste, which will eventually be vitrified, is at present stored at Sellafield in sixteen double-skinned, cooled, stainless steel tanks surrounded by concrete 2 m thick. This material must be isolated from the environment, essentially for ever.

The organic (TBP/OK) phase is treated with ferrous sulphamate to separate uranium and plutonium. The plutonium is converted to the oxide and stored, whereas the uranium is recycled. It would, however, be naive to assume that the plutonium separated is all for civil use; some plutonium, notably that from the Chapelcross power station, is for weapons production. The process for oxide fuel (AGR/PWR) is very similar to the above except for the initial decladding, which is to be replaced with a total dissolution regime.

It will be clear from the description above that there are radioactive wastes that need control from nearly all the stages of the nuclear fuel cycle. Their impact on workers and the general population used to be considered when estimating risk–benefit balances, e.g. uranium miners are exposed to considerable risk (from radon and its daughters) but are excluded from this balance equation, and again, for instance, the people of Ireland are affected by the discharges from Sellafield but reap no benefit. The most important state of the fuel cycle from the point of view of risk to the population from discharges is reprocessing and therefore emphasis will be concentrated on this part of the process.

The most important aspect of waste management from this part of the process is that the UK still has no definite long-term policy for handling the radioactive waste generated. This is an indictment of the nuclear industry, which is now thirty years old. Investment on a massive scale is being encouraged at Sellafield but with only an apparently hazy idea of the ultimate fate of all types of waste; namely, low-level waste (which is currently buried at BNFL's dump site at Drigg or discharged into the Irish Sea), intermediate-level waste (which is stored in drums pending some decision on a site for a repository) and high-level waste (which is stored in tanks at Sellafield at present but will be 'vitrified' when the plant is available, and stored 'somewhere'. In this context, it is of interest to recall some oft-quoted recommendations of the Royal Commission on Environmental Pollution made in 1976:

> There should be no commitment to a large programme of nuclear fission power until it has been demonstrated beyond reasonable doubt that a method exists to ensure the safe containment of long-lived light radioactive waste for the indefinite future.
>
> There is a lack of clearly formulated policy for the disposal of intermediate level solid waste at nuclear stations. The policy of accumulating more highly active solid wastes at AEA and BNFL sites with a view to eventual ocean disposal appears inadequate. Such disposal may prove unacceptable and the possible future requirements again point to the need for a national disposal facility.

This lack of a formulated policy is amazingly still evident thirteen years later! For instance NIREX talks of a deep repository not being available for at least ten years. This is at a time when a number of our elderly Magnox stations are to be decommissioned, we are to embark on a four-station (at least) PWR programme, and a number of nuclear submarines have to be disposed of. There have been a number of U-turns, e.g. the abandonment of Killingholme, Bradwell, Elstow and Welbeck as dump sites, after they had been selected by NIREX, and the decision not to sea dump, which were forced on the government by public pressure. However, misjudgements of the extent of public feeling which preceded these changes in policy are disturbing. The Radioactive Waste Management Advisory Committee (RWMAC) complains that 'there remain a significant number of issues on which policy remains disappointingly confused or deficient'. It also pointed out in 1988 that:

> there is enormous uncertainty whether a deep disposal site will be found, assessed and in place by 2005 as planned. In addition, there is doubt about the eventual disposal route for large scale decommissioning items and confusion over the degree to which intermediate level waste (ILW) conditioning might be allowed to proceed without foreclosing disposal and treatment options ... On top of all these deficiencies in the strategic planning, we are continually met with pleas about the severe limitation upon resources for the radiochemical inspectorate arm of Her Majesty's Inspectorate of Pollution (HMIP) which is required to administer the present system of administrative and technical controls.

RWMAC also points out that the situation is little better in other OECD countries, with the notable exception of Sweden which is phasing out nuclear power, as illustrated in Table 9.2.

The confusion in the civil service and the public over who is responsible for what in waste management is well illustrated by Figure 9.3, which is reproduced from the 9th annual report of RWMAC. The state of the art in radioactive waste management is put into perspective by a somewhat rueful conclusion from the 1988 RWMAC report:

> Broadly speaking it seems likely that technical issues will be less important than social and political considerations. It is possible to design suitable solutions to the technical problems of LLW and ILW. But the social and political issues might pose the greater problems.

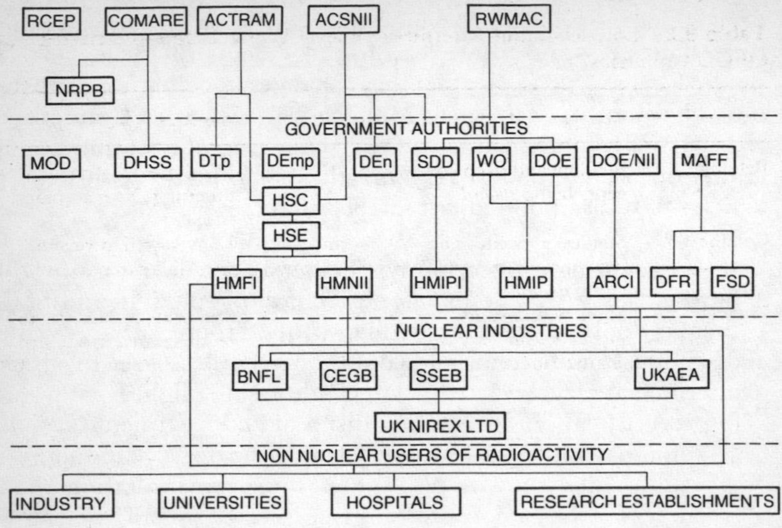

Figure 9.3 UK Agencies Involved in Study and Control of Radioactive Waste.

LIST OF INITIALS

ACSNI	Advisory Committee on the Safety of Nuclear Installations
ACTRAM	Advisory Committee on the Transport of Radioactive Material
ARCI	Alkali and Radiochemical Inspectorate
BNFL	British Nuclear Fuels plc
CEGB	Central Electricity Generating Board
CMO	Chief Medical Officer
COMARE	Committee on Medical Aspects of Radiation in the Environment
DEmp	Department of Employment
DEn	Department of Energy
DFR	Directorate of Fisheries Research
DHSS	Department of Health and Social Security
DOE	Department of the Environment
DOE/NI	Department of the Environment for Northern Ireland
DTp	Department of Transport
FSD	Food Science Division
HMFI	HM Factories Inspectorate
HMIP	HM Inspectorate of Pollution
HMIPI	HM Industrial Pollution Inspectorate for Scotland
HMNII	HM Nuclear Installations Inspectorate
HSC	Health and Safety Commission
HSE	Health and Safety Executive
IAEA	International Atomic Energy Agency
MAFF	Ministry of Agriculture, Fisheries and Food
NRPB	National Radiological Protection Board
RCEP	Royal Commission on Environmental Pollution
RWMAC	Radioactive Waste Management Advisory Committee
SDD	Scottish Development Department
SOS	Secretary of State
SSEB	South of Scotland Electricity Board
UKAEA	United Kingdom Atomic Energy Authority
WO	Welsh Office

Table 9.2 Low-level and Intermediate level Waste Disposal Activities in OECD Countries

Country	*Concept*	*Host rock*	*Comments*
Belgium	Shallow and deep concepts being *studied*	Clay	LLW previously to sea dumping. Now stored at Mol.
Canada	Shallow repository or near surface caverns being *studied*	Clay and sandstone caverns	LLW stored at reactor sites and two purpose built stores.
France	Shallow repository incorporating extensive engineering *implemented*	Clay	Centre de la Manche to be succeeded by Centre de l'Aube around 1990–1.
West Germany	Deep disposal sites being *investigated*	Salt domes. Abandoned iron ore mine	Asse mine was used for disposal of LLW and ILW from 1967 to 1978. Konrad mine being developed for non-HGW. Gorleben being investigated for all wastes.
Finland	Underground repository being *implemented*	Granite	Similar to Swedish facility.
Japan	Near surface disposal in engineered concrete bunkers being *implemented*	Unknown	Shimokita disposal centre to be operational around 1991.
Sweden	Deep disposal in caverns 50 m under the sea bed being *implemented*	Granite	The SFR facility at Forsmark is now operational.
Switzerland	Caverns built into mountain side being *investigated*	Marl/granite	Swiss Type B (LLW ILW) repository will be under several hundred metres of overburden.
USA	Shallow land burial in excavated trenches *implemented*	Various	LLW Policy Amendment Act 1985 demands that each state makes arrangements for a disposal site, singly or in a consortium.
United Kingdom	Shallow land burial *in progress*. Shallow engineered trenches being *developed*. Deep disposal being *studied*	Various for deep disposal	Drigg and Dounreay sites in operation. Concepts being investigated for deep disposal site. Available around 2005.

Low-level Waste

In spite of this administrative chaos, discharges of low-level waste have proceeded. Let us now look more closely at the fate of discharges from Sellafield because, up until now, it has been felt that LLW could be judiciously released to the environment on the 'dilute and disperse' principle, rather than stored. This LLW has therefore had the greatest direct impact on members of the public but, incredibly, after more than thirty years of this policy, the government has stated this 'should not be a general principle for waste management'.

Solid waste

This waste, which is characterized by its large volume and low level of radioactivity, is defined by NIREX and RWMAC as having less than 4 GBq/tonne of alpha activity or 12 GBq/tonne of beta/gamma generated at all stages of the nuclear fuel cycle. Large volumes are produced in the UK each year, and its

Table 9.3 Radioactivity in the Drigg Stream (Bq/litre)

Year	*U*	*Sr-90*	*Pu*	*H-3*
1977	2.6	7	0.07	
1978	1.9	18	0.12	8000
1979	2.9	24	0.07	7000
1980	1.7	16	0.11	7500
1981	2.0	22	0.22	8200
1982	2.3	29	0.40	15000
1983	2.9	27	0.30	15000
1984	2.9	39	0.20	44500
1985	3.1	33	0.20	26000
1986	4.3	33	0.61	45000
1987	3.1	45	0.30	30500

management and disposal are a matter of some contention. Solid waste of this type from Sellafield is taken to the Drigg site, 6 km south of Sellafield, for burial, currently in long, concrete-lined trenches. This waste has an alpha activity of no more than 740 kBq/m^3 and beta activity of no more than 2220 MBq/m^3. There is a similar smaller site at Dounreay.

Typical constituents of waste disposed of at Drigg are contaminated paper, cardboard, plastic, protective clothing, electric cabling, scrap metal, excavation spoil, filters, resins, catalysts and general laboratory waste. Disposal trenches on the Drigg site are drained into a stream that runs off the site and discharges into the intertidal region of the River Irt in the Ravensglass Estuary. This stream, the Ravensglass Estuary and milk from a local farm are all monitored. The concentration of radioactivity in the stream is gradually increasing as usage of the dump increases (Table 9.3). There are plans to improve the site and some have been implemented, but improvements in waste disposal were also recommended in the Royal Commission (Flowers) Report, twelve years ago, which argued that: 'The burial of low-level solid wastes at Drigg site is satisfactory for the present but in the longer term such wastes should preferably be taken to a national disposal facility'.

Although the levels of radioactivity leaching from the site are still quite small – BNFL estimate consumption of 300 litres of the stream would be necessary for a radiation dose of 0.1 mSv – these are increasing as the capacity of the facility is reached. The Drigg site has recently been the subject of scrutiny by the Government Environment Committee as a live issue, because of more waste generation from the proposed PWR programme, was mentioned at the Hinkley Point Inquiry (1989). The government seem to have accepted that Drigg is likely to be full by 2010 and that this could only be achieved by compaction of waste received there. However, BNFL have been committed to a site improvement plan including more restrictive authorizations. This makes the search for a properly engineered 'National Disposal Site' imperative.

Low-level waste that was too active for the Drigg site was set in concrete in steel drums and, between 1949 and 1982, dumped in the North Atlantic. The location of dumping sites varied over the years (Figure 9.4) but the final site chosen was about 500 miles southwest of Lands End, where the sea was about 4 km deep. Following considerable opposition from pressure groups and the

National Union of Seamen, the dumping ceased in 1982. The Nuclear Energy Agency (NEA) of the OECD had supervised the dumping but the amount dumped had been quite substantial (Table 9.4) and in May 1988, the UK government officially abandoned sea dumping. However, in reviewing the options for waste management in their 9th annual report, RWMAC clearly approved of sea-bed disposal.

Liquid waste

This waste arises at most stages of the nuclear fuel cycle, particularly during reprocessing but also from the power stations themselves. At Magnox and AGR power stations, liquid wastes contain small amounts of activation and fission products. The former come from activation of impurities in the carbon dioxide cooling gas or materials used in the construction of the reactor. Carbon, for instance, either in the graphite moderator or carbon dioxide gas coolant, can capture two neutrons and become C-14, a beta emitter of long half life. Activation of trace metals in the steel of the reactor produces Co-60 and isotopes of manganese, chromium and iron. Corrosion of the fuel cans in the cooling pond or even leaking fuel cladding can lead to fission-product contamination of the ponds or even the coolant gas. Although every effort

Table 9.4 Quantities (TBq) of Waste Dumped in the north-east Atlantic, 1949–82

	Alpha	*Beta/gamma*	^3H	*Countries supplying waste*
1949–55	3.3	9.0	0	UK
1956–60	71.5	90.1	0	UK, Belgium
1961–65	56.9	1411	0	UK, Belgium
1966–70	65.9	4750	0	UK, Belgium, France, Germany, Netherlands, Sweden, Switzerland, Italy
1971–75	118	7480	1100	UK, Belgium, Netherlands, Switzerland
1976–80	229	17500	8570	UK, Belgium, Netherlands, Switzerland
1981–82	134	10400	5700	UK, Belgium, Netherlands, Switzerland
1983 – no more dumping				

Note α activities include radium-226 and β/γ include tritium unless separate figure given.

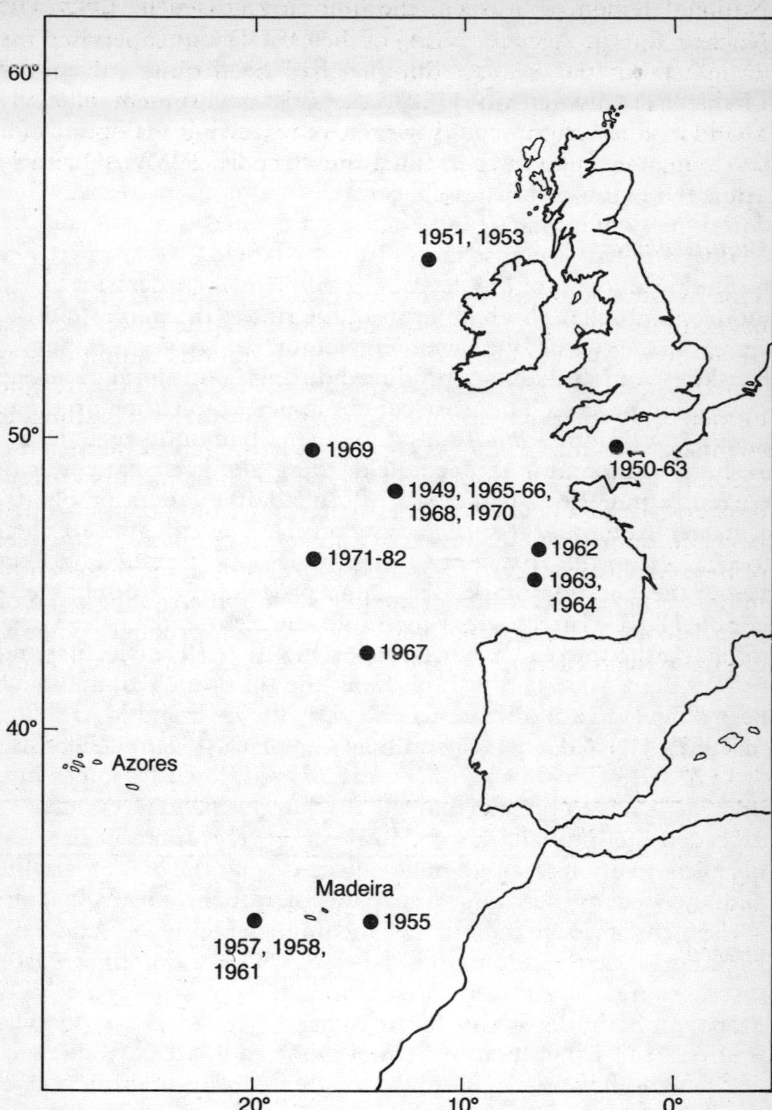

Figure 9.4 Solid Radioactive Waste Disposal Sites in North European Waters.

is made to clean these by filtration and ion exchange, small amounts do slip through. In the UK, low-level liquid wastes from nuclear power stations are discharged into the sea (except for Trawsfynydd station whose effluent goes into an artificial lake). The levels discharged from power stations are not very great and this can be seen in the current doses to critical groups given in Table 9.5. However, these local releases may increase when decommissioning starts and their significance is small only by comparison with the discharges from Sellafield.

The production of low-level wastes during reprocessing is far more of a problem, both in terms of the risk to the public and the scale of the waste management required. Large volumes of low-level liquid wastes are produced during pond storage and the separation process. These are contaminated with fission products and, perhaps more importantly, with small amounts of actinides such as plutonium. At Sellafield these liquids (effluent) are currently put through the SIXEP and Salt Evaporator Plants, which substantially reduce activity levels. The effluents are then discharged to the Irish Sea through pipes that terminate 2.1 km below the low tide mark. The same procedure for discharge is adopted at Dounreay via a pipe into the Pentland Firth. These discharges to the sea have been occurring since 1952 (the first and second pipes were laid in June 1950, the third in 1976) and have been subject to authorizations issued by the DoE and MAFF. Up to 1987, 41,104 TBq (1.11-million Ci) of Cs-137 and 683 TBq (18,470 Ci) of Pu-239 had been discharged down the pipes into the sea. It has to be said that the discharges have never actually exceeded the limits imposed. However, several times in the past the authorizations seem to have reflected what the operator of the plant needed, or the plant was capable of, rather than an objective assessment of risk to members of the public (Figures 9.5 and 9.6). In order to test the dilute-and-disperse policy for dumping waste, the discharge levels were 'deliberately' increased in the early 1950s to obtain detectable levels in fish, seaweed and sand. The results of this experiment were first reported to a UN Conference in 1955 and have since hung like an albatross around the neck of the perpetrator, John Dunster (then health physicist with the UKAEA, later director of the NRPB). Throughout the years the plant has been required to operate the BPM principle (best practible means) in reducing discharges. For contravening this principle, BNFL was fined in 1983 after it allowed a substantial

Table 9.5 Summarised Estimates of Public Radiation Exposure from Discharges of Liquid Radioactive Waste in the UK, 1988 (Nuclear Power Stations Operated by the Elecricity Boards)

Berkeley and Oldbury	Fish and shellfish consumption	Local fishing community	<0.01
	External		
Bradwell	External	Houseboat dwellers	0.02
Dungeness	External	Bait diggers	<0.01
	Fish and shellfish consumption		
Hartlepool	Fish and shellfish consumption	Local fishing community	<0.01
	External	Coal collectors	<0.002
Heysham	Fish and shellfish consumption	Local fishing community	0.07 (0.11)
	External		<0.1
Hinkley Point	Fish and shellfish consumption	Local fishing community	<0.01
	External		
Hunterston	Fish and shellfish consumption	Local fishing community	0.02
	External		
Sizewell	Fish and shellfish consumption	Local fishing community	<0.003
	External		
Torness	Fish and shellfish consumption	Local fishing community	<0.005
	External		
Trawsfynydd	Fish consumption	Local fishing community	0.07
	External		
Wylfa	Fish and shellfish consumption	Local fishing community	<0.02
	External		

Note * The first value is based on the gut transfer factor for plutonium and americium of 0.0002; the value using a factor of 0.0005 follows in parentheses.

amount of material to be discharged that eventually found its way back on to the West Cumbrian beaches; a 40-km stretch of beach was declared 'not recommended for use' for a time.

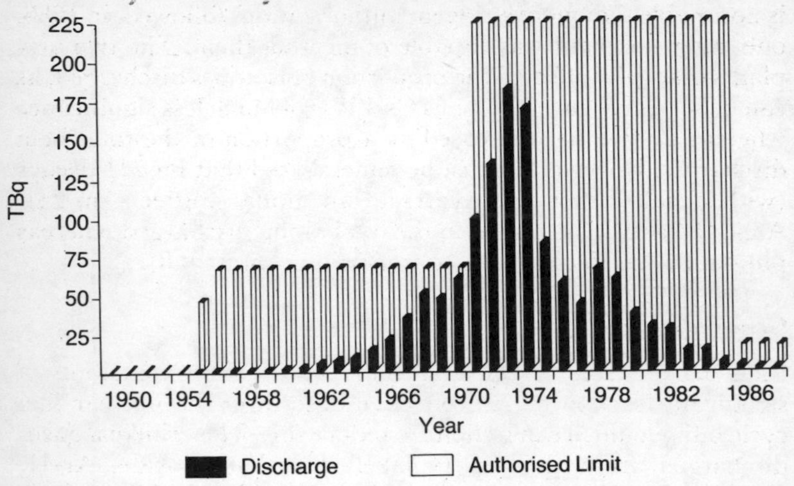

Figure 9.5 Total Alpha Discharges from Sellafield, 1950–88.

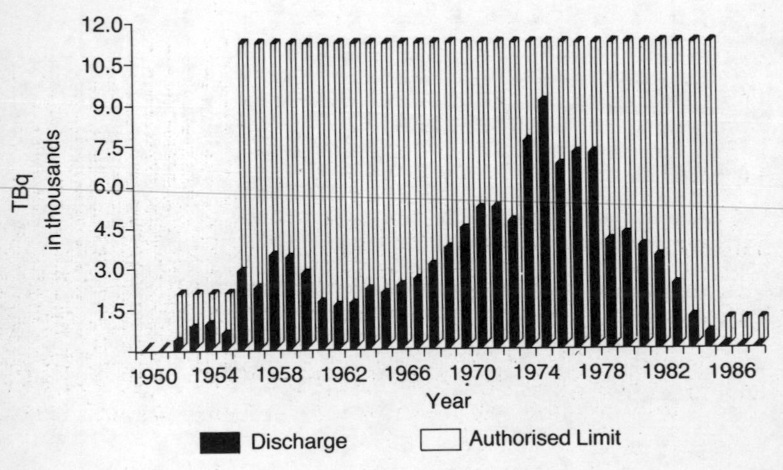

Figure 9.6 Total Beta Discharges from Sellafield, 1950–88.

The discharges by year from 1952 are given in Figure 9.7 for selected radionuclides and also for total alpha and total beta in Figures 9.5 and 9.6. The marked effect of the introduction of SIXEP and the Salt Evaporator in 1985 can be seen. However, it is noticeable that more stringent authorizations followed, in 1986, only after the plant was capable of meeting them! The two new plants markedly affected the proportion of isotopes discharged, as can be seen in Figure 9.8. Thus Cs-137 is of much less significance whereas Pu-241 has increased as a proportion of the total beta discharges. However, it must be remembered that Pu-241 decays (with a half life of 13.5 years) to an alpha emitter, Am-241. Am-241 has also increased in the total alpha discharged whereas plutonium has decreased as a proportion (Figure 9.8).

Gaseous waste

Radioactive gases released into the atmosphere contribute a significant fraction of the collective dose from the nuclear fuel cycle but not much can be done to reduce this. The principal gases discharged are tritium, C-14 labelled carbon dioxide, Ar-41, Kr-85 and I-129, with short-lived Xe-135 and N-16 adding a little to the local dose. There is some leak of fission-product gases from corroded fuel cans in cooling ponds but the main gases discharged from nuclear power stations are activation products. C-14 and

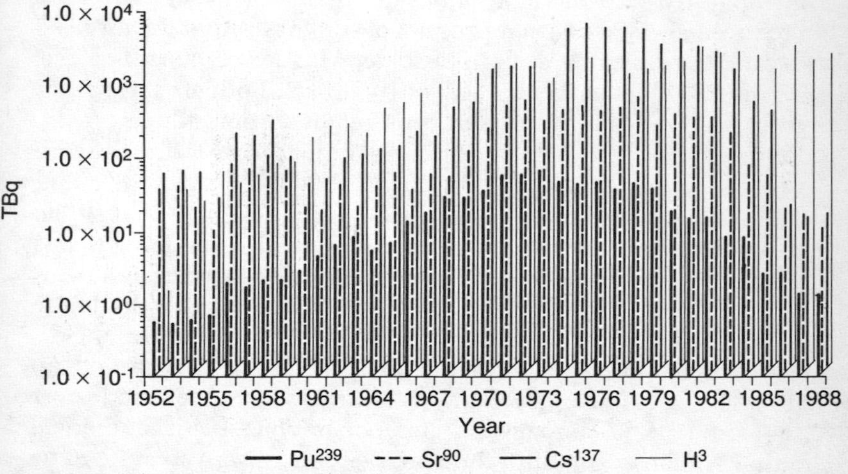

Figure 9.7 Discharges from Sellafield into the Irish Sea (TBq/year).

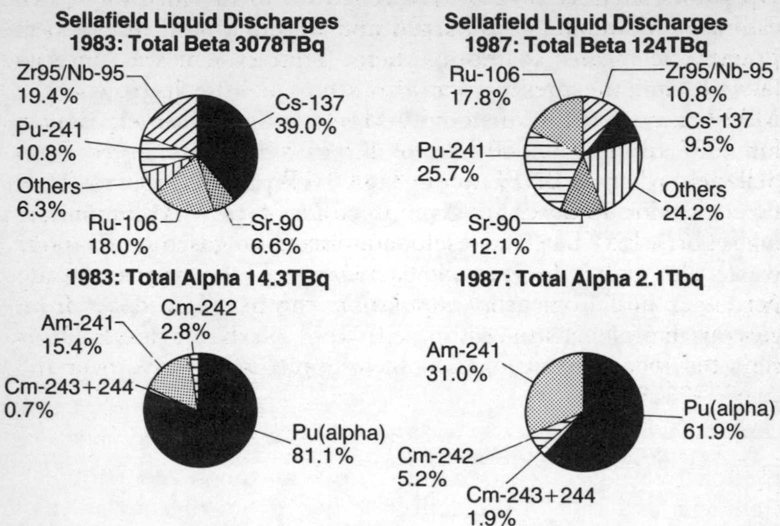

Figure 9.8 Beta and Alpha Constituents of Sellafield Liquid Discharges, 1983 and 1987.

H-3, for example, are both produced in the reactor by neutron capture. These are both low-energy beta emitters but can be incorporated into the food chain in organic molecules. Ar-41 is produced as an activation product of stable Ar-40 in the air used to cool the pressure vessels in the older Magnox reactors. Kr-85 is a fission product with a half life of 10.5 years. Both are inert gases and therefore difficult to scrub or remove from effluent gases. Releases of 11 TBq/MWe-year of A-41 are typical whereas Sellafield released about 34,000 TBq (91,900 Ci) of Kr-85 and 2400 TBq (65,000 Ci) of Ar-41 in 1987. Only since 1 January 1988 have numerical authorization limits been in force for gaseous discharges and these vary, for Sellafield, with stack height. For Sellafield, the authorization for Kr-85 is specific for discharge points higher than 70 m and is 100,000 TBq/year. Sellafield's aerial discharges are dominated by Kr-85 as are those from Cap de la Hague (Figure 9.9). Various estimates have been made over the years of its impact on the population. Until numerical authorizations came into force in 1988, there was no systematic environmental monitoring for radioactive gases such as Kr-85.

Operators such as BNFL were relied on to produce their own estimates of amounts discharged and were required to use 'best practicable means' to reduce them. That system was unsatisfactory, and the present one also seems a little hazy, e.g. the MAFF Terrestrial Radioactivity Monitoring Programme Report for 1987 records 40,000 TBq of Kr-85 being discharged from Sellafield whereas BNFL quote 34,000 TBq in their report! Also, no critical group doses are given in either of the MAFF monitoring reports. BNFL itself does no monitoring for gaseous or vapour waste, although it does sample particulates in the air near the site perimeter and in nearby population centres. The doses from gaseous emissions are 'estimated' from discharge levels. It is reported that they 'would not have exceeded 0.1 mSv over the year' (1987).

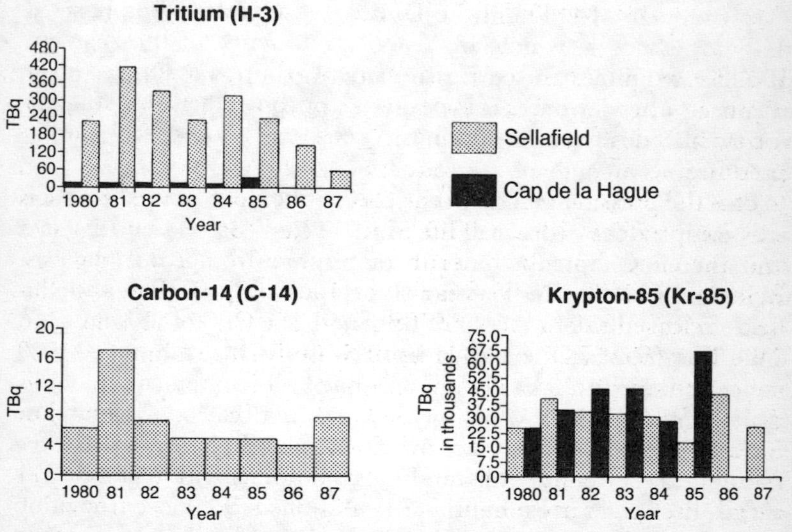

Figure 9.9 Radionuclides Discharged from Nuclear Fuel Reprocessing Plants as Airborne Effluents, Sellafield (1980–7) and Cap de la Hague (1980–5).

Doses to Critical Groups

The control of the discharge of radioactive waste to the environment in the UK is based on an assessment (it can be no more than that) of the dose to what is termed 'a representative member of a critical group'. The critical group for any site is a group of people who, because of their dietary habits or domicile, are thought to be likely to receive the highest radiation dose, the principle being that if this group is protected, all other groups will be overprotected. This policy does, however, depend on continuous 'habit surveys' and a thorough knowledge of metabolic and environmental pathways. In many respects, the policy does involve some informal prediction of environmental pathways and this has been found wanting on some occasions. The most notable of these faulty predictions was the anticipations that actinides, particularly plutonium, released into the Irish Sea from Sellafield would remain bound to the bottom sediments. This proved to be not entirely the case and sea-suspended actinides are blown in spray on to the beaches where the material dries out and is blown inland (Figure 9.10).

In practice, MAFF and the DoE have separately emphasized that target doses, which are usually one-tenth of the relevant public dose limit, are used for the critical group to ensure a degree of conservative latitude. It is perhaps appropriate at this stage to review the dose limits that have been used to control public exposure.

The ICRP recommended that the public dose limit should be one-tenth of the worker dose limit. The reasons for this are explained in Chapter 3. This philosophy was accepted in the UK and built into the Radioactive Substances Act of 1960 and the Nuclear Installations Act of 1965. Since the worker annual dose limit was 50 mSv, the public limit was set at 5 mSv per year. Subsequently, in 1985, the ICRP made the statement that the limit should be 1 mSv per year with an occasional excursion allowable to 5 mSv per year, provided a lifetime dose of 70 mSv was not exceeded. Notwithstanding an endorsement by the NRPB of this, the Ionizing Radiations Regulations (1985) gave the limit as 5 mSv per year. However, the DoE and MAFF have stated that they take into account the objectives of radioactive waste management as set out in government White Papers. In 1986, the government revised these objectives in its response to the

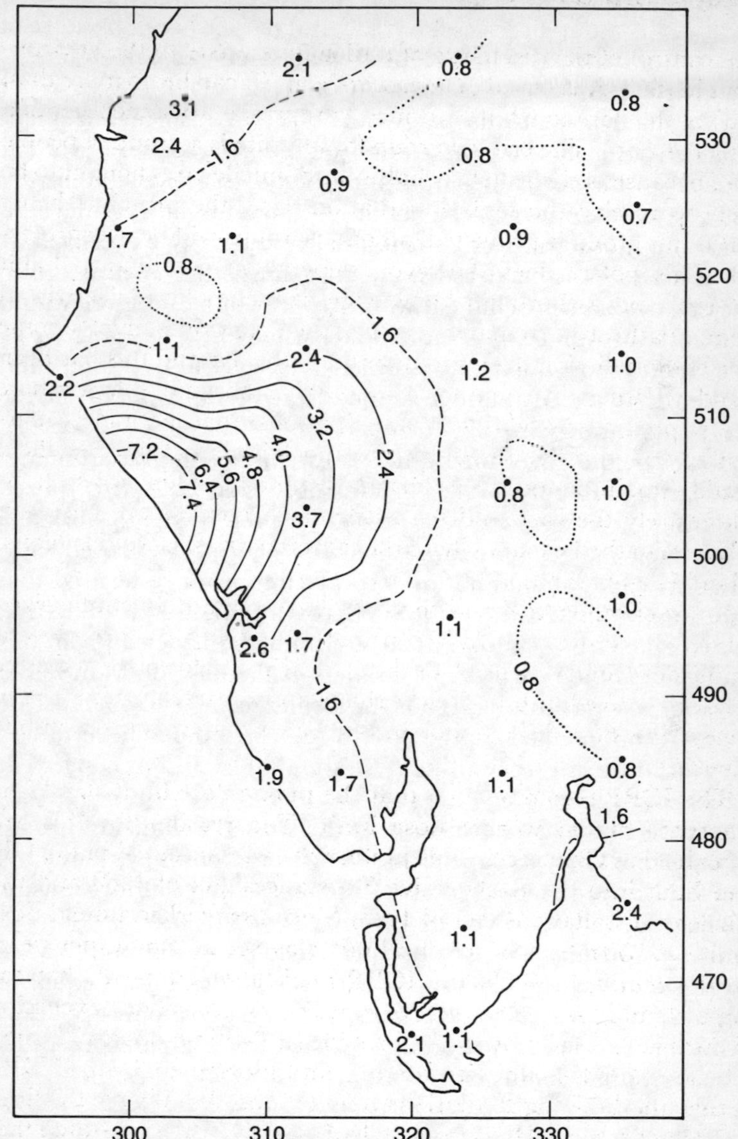

Figure 9.10 Plutonium-239 Levels (Relative to Fallout) in Cumbrian Grassland, 1978.

Environment Committees Report on Radioactive Waste as a result of Euratom directives (80/836 and 84/466) and, because of advice from the NRPB, accepted the ICRP recommendation. Since then (1987) both the RWMAC and the NRPB have advised that for all nuclear site discharges an objective waste management practice should be that the committed effective dose equivalent (CEDE) to the critical group should be no greater than 0.5 mSv per year. This appears to be no more than a target, i.e. a level to aim at, but not so restrictive as to involve a penalty if exceeded. If this were to be considered as a dose limit, the critical group assessed doses near Sellafield would approach it and there would be a certainty that some individuals would be exceeding it. In their 1988 report, BNFL welcomed this level as a 'target' and recalled that it had made a similar 'statement of intent' at the Windscale Inquiry in 1978.

However, the government, in its response to the Environment Committee's Report (1986) confusingly said that it required NIREX, when designing repositories for solid waste, to work to a target such that the 'maximum risk to the most exposed individual in a year should not under any circumstances exceed that associated with a dose of 0.1 mSv: that is a chance of death to that individual of 1 in a million'. It is not immediately clear from this whether it is the risk of 1 in 1 million or the dose of 0.1 mSv per year that is to be limiting. If it is the former, using current risk rate figures recommended by the NRPB (4.5 in 100 per sievert – see Chapter 3) the target limiting dose should be 0.022 mSv per year (22 μSv per year).

Members of critical groups near the sites at Sellafield, Springfields, Capenhurst, Chapelcross, Winfrith, Dounreay, Heysham and Trawsfynydd would receive doses exceeding this target using MAFF 1988 figures. Even if 0.1 mSv per year is used as a target, critical groups near Sellafield, Springfield, Capenhurst, Chapelcross, Dounreay and Heysham will receive doses that equal or exceed it. However, it could be that this target dose only applies to waste repositories and the target is five times greater for other critical groups. Confusion reigns!

Monitoring Around Nuclear Sites and Pathways to Man

Sites associated with the nuclear fuel cycle in the UK are now extensively and systematically monitored. Monitoring in areas around non-nuclear energy sites is virtually non-existent. Thus, for the majority of small users of radioactive materials there are virtually no checks of how and to what extent they are polluting the environment. However, because of the immense pollution problem that nuclear fuel cycle sites could pose, it is probably best to concentrate meagrely funded monitoring effort where it is needed most.

The main purposes of environmental monitoring are to assess public exposure, to give some warning of unexpectedly high discharges and to define critical group pathways. The environmental monitoring around a site, such as Sellafield, is carried out largely by two organizations, BNFL and MAFF. The DoE do some monitoring (spot checks) on water and sediments but the main monitoring effort is done by the site operator (BNFL), with MAFF doing routine, systematic checks at similar locations. MAFF's monitoring covers not only the aquatic environment but also, since 1986, food pathways via deposition on land. Results from all monitoring programmes are published separately, annually, although no organization at present seems to have responsibility for monitoring for Kr-85 in the air.

No matter how carefully sampling is done, there will, it seems, always be occasions when a pathway to man is uncovered that had not been considered before. This is, of course, becoming increasingly less likely, but with a complex environment such as that in the northeastern Irish Sea, the 'critical group' concept is especially fragile. For instance, DoE revealed unexpectedly high levels of Am-241 in the Esk Estuary (south of Sellafield) in 1989. These pose no immediate significant hazard but do emphasize the need for constant vigilance.

Apart from regular routine monitoring of food, and the environment, a great deal of research has been done to plot the spread of radioactivity from the Sellafield site. The reports make interesting reading: Fission products (e.g. Cs-137) spread out from the site in regular iso-radioactivity levels, as do levels of the actinides (e.g. Pu-239). This alone was a little unexpected in that although caesium was likely to be soluble and therefore dispersed in the sea, about 95 per cent of it remains in solution. It was readily bioaccumulated and therefore thought to be available only from the aquatic food chain. Also, it was thought that actinides such as Pu, Am would be bound (95 per cent) to sea bottom sediments and would stay there, affecting only bottom feeders. It was found, however, that this binding was incomplete and blown sediment in intertidal regions and suspended particles from seaspray presented an additional pathway to man (Figure 9.11). The spread on land is much less marked with Cs-137 which may be more influenced by atmospheric discharges from Sellafield including the release of 22 TBq in the fire in 1957.

The pathways and therefore dose to man of this material blown and deposited inland are largely via inhalation, with virtually no contribution from material absorbed from contaminated, locally grown food. This has been confirmed recently by monitoring of duplicate diets published by MAFF in 1986 and 1987. However, the inhalation of plutonium- and amercium-contaminated house dust may yet prove to be the most important source of radiation dose from the Sellafield discharges. This problem was first identified in 1976 just before the planning inquiry for THORP in 1978. Lord Parker, the Inspector, charged the NRPB with the task of measuring samples of house dust from the Ravensglass area. This they did, and their survey was followed by a much more extensive one by Imperial College. The results of both showed considerable contamination by plutonium and americium (about 50 Bq/kg), viz., plutonium-239 was increased by a factor of about 250 and americium-241 by a factor of about 6000

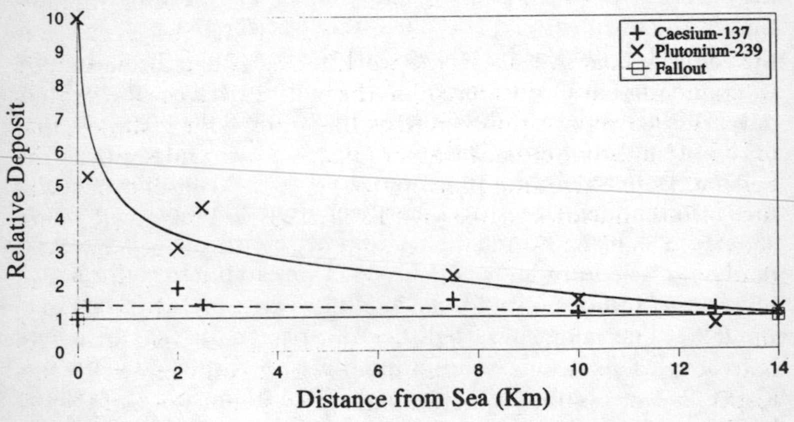

Figure 9.11 Deposition of Plutonium-239 and Caesium-137 Relative to Estimated Weapon Fallout at St Bees/Ennerdale.

when compared with house dust from the southeast of England. However, considerable controversy still exists over the risk posed to people from this dust. The contentious issues revolve around the respirable fraction of the dust and the relevant dose delivered to the bronchial epithelium (in the lung). However, whatever the finer points of the argument, the existence of increased levels of actinides of this magnitude must be a cause of some concern. This is a point of particular anxiety because, inexplicably, actinides were not included in analyses of environmental material until 1973, even though by then 500 TBq of alpha activity had been dumped into the sea, and the inhalation hazard from resuspended sediments was not explained until 1976.

From the published results concerning plutonium that has spread inland from the Sellafield discharge pipe it is clear that a coastal strip about 5 km wide and about 40 km long has been contaminated with at least 100 GBq of Pu-239. In addition, as discharges have dropped by about a factor of ten in the last seven years, estuaries such as Ravensglass are now thought to be net exporters of plutonium.

The spread of radioactivity discharged from Sellafield to the sea has been fairly well explored both by MAFF and others. Its spread throughout the seas around Western Europe is shown in Figure 9.12 as Bq of Cs-137/litre of sea water. Bearing in mind the number of nuclear sites that discharge waste into these coastal waters (Figure 9.13) it is obvious that Sellafield's discharges have by far the greatest influence. It is often said that the Irish Sea is the most radioactive sea in the world. This is true if man-made radionuclides are considered in the Cumbrian coastal waters only. A 1989 report from UKAEA has shown, for instance, that only in Cumbrian coastal waters (10 km north and south of the Sellafield site, extending 2 km out to sea) is the overall concentration of radionuclides increased (Table 9.6). The artificial radioactivity of the Irish Sea will be dominated by those radionuclides that remain in solution, i.e. isotopes such as Cs-137, Sr-90 and H-3. However, it must be remembered that even the relatively small amount of plutonium that does not become bound to bottom sediments may be a significant risk contributor. In this respect it is noteworthy that although the concentration of Cs-137 has been studied extensively both in the Irish Sea and elsewhere, there are few such measurements for H-3, Sr-90, Pu-238, Pu-241 and Am-241. Caesium has been thought of as a tracer material

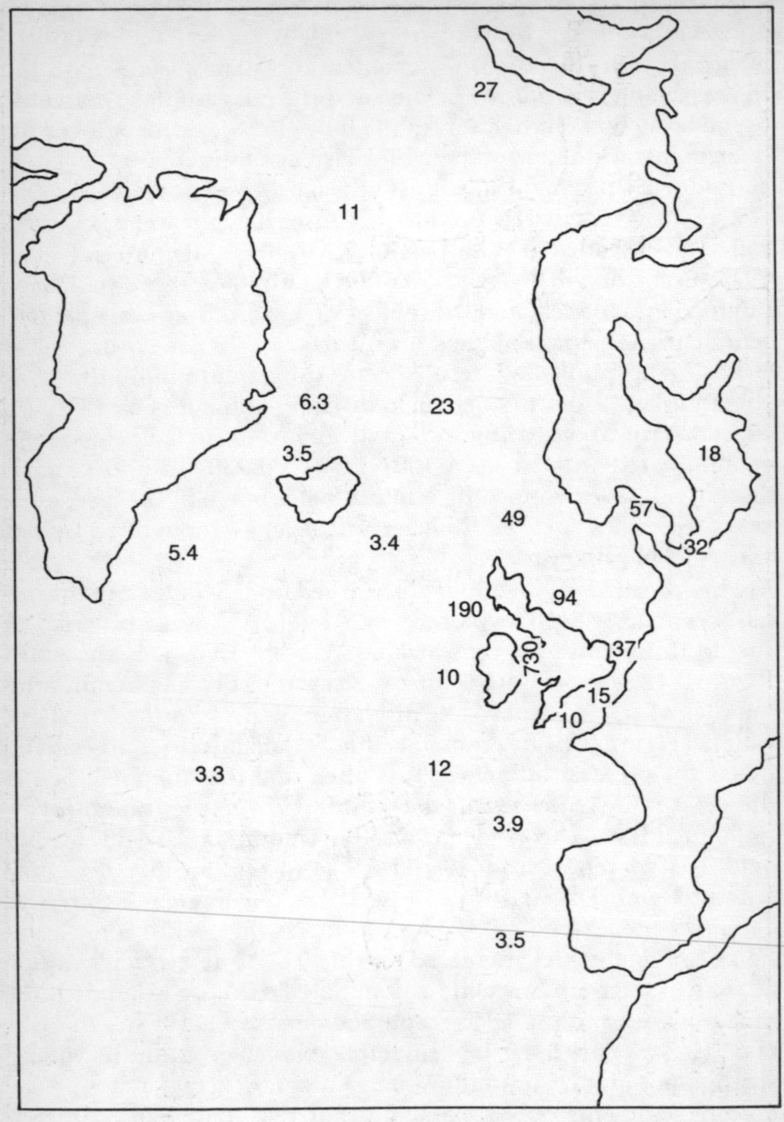

Figure 9.12 Average Concentrations of Cs-137 in Western European Waters, 1980–5 (Bq/m^3).

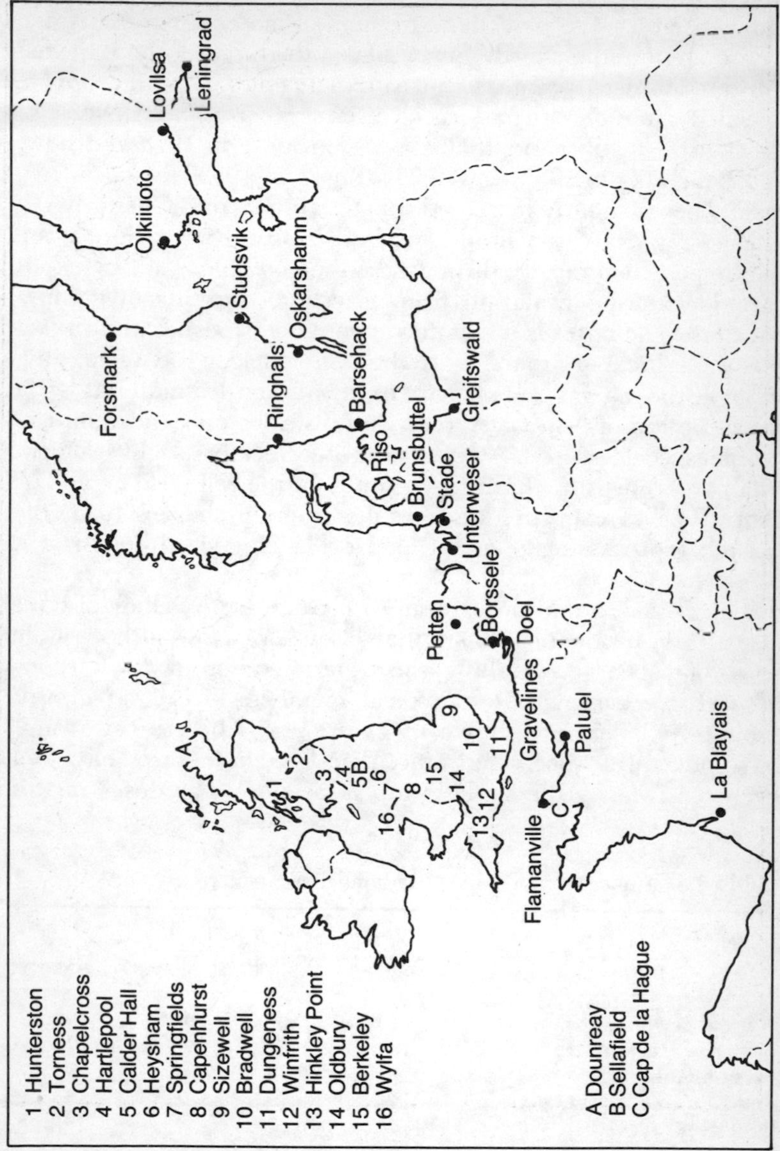

Figure 9.13 Nuclear Plants Discharging into the Coastal Waters of Northern Europe.

that represents the behaviour of all isotopes remaining in solution. It has been used to plot the movement of material via the North Channel into the North Sea and southwards. Although vast dilution has occurred, these measurements in the North Sea have revealed the different pattern of discharges from Sellafield, e.g. 15–40 mBq/kg of sea in 1967, 50–250 mBq/kg in 1978 and now, 10–50 mBq/kg in 1987, and 5–25 mBq/kg in 1988 (the latter two results being slightly increased due to fall-out from Chernobyl).

The impact of the liquid discharges from Sellafield on the public has been mostly through the seafood chain and by exposure to external radiation from intertidal deposits of gamma emitters. The latter is a significant route of exposure and in fact sets the critical group for more than one nuclear site. Generally, though, the maximum exposure to the discharged material is via consumption of fish, crustaceans (crabs, lobsters) and molluscs (winkles, cockles, etc.). Avid consumers of locally caught sea food, therefore, constitute the critical groups for the Sellafield area. The same type of exposure also results from discharges from the French reprocessing plant at Cap de la Hague, although at a somewhat lower level.

The critical group doses from discharges from Sellafield have also always been much higher than elsewhere. Now, although the levels of discharge are falling, the radiation doses may be a legacy of past excesses. Doses to typical members of critical groups around Sellafield are shown in Figure 9.14 (the critical groups have not always been the same, therefore these doses have not been received by the same group of people). The doses in this

Table 9.6 Concentrations of All Radionuclides in Various Sea Waters

Type of sea water		*Concentration (Bq/kg of water)* *Natural*	*Fallout*	*Sellafield*	*Total*
Normal ocean		12.6	1	0	13.6
Eastern Mediterranean		13.6	1	0	14.6
Persian Gulf		21	1	0	22
Red Sea		14	1	0	15
Dead Sea		177	1	0	178
Baltic Sea		4.0	1	0	5
Irish Sea	1987	12.0	1	0.7	13.7
Cumbrian coastal waters	1987	11.5	1	59.4	72

figure are abstracted from a number of reports (mostly MAFF) and are the highest that have been estimated for the relevant critical groups. Thus some are for external exposure, some via the consumption of seafood. The figures for Trawsfynydd power station are included as these have tended to be the highest for sites other than BNFL plant. The discharges from Trawsfynydd go into an inland lake that drains to the sea. The critical groups here are local fishermen who not only eat fish from the lake but also receive external exposure at the lakeside. The few results from the COGEMA reprocessing plant at Cap de la Hague are included again for comparison. These doses would be received after 300 hours per year occupancy over contaminated silt in the area.

One important, though somewhat contentious factor in the estimation of critical group doses from discharges is the amount of material (particularly plutonium and other actinides) that is absorbed from food, the gut transfer factor. For several years MAFF has reported two critical group doses because it (and other organizations) has been uncertain as to which factor to use (Table 9.5 only quotes the dose from the highest, most conservtive in protection terms). The original figure for actinides was given in 1959 by ICRP (in Publication 2) and endorsed in 1980 in its Publication 30. This figure indicated that 10^{-4}, i.e. one ten-thousandth, or 0.001 per cent, of plutonium in the diet was

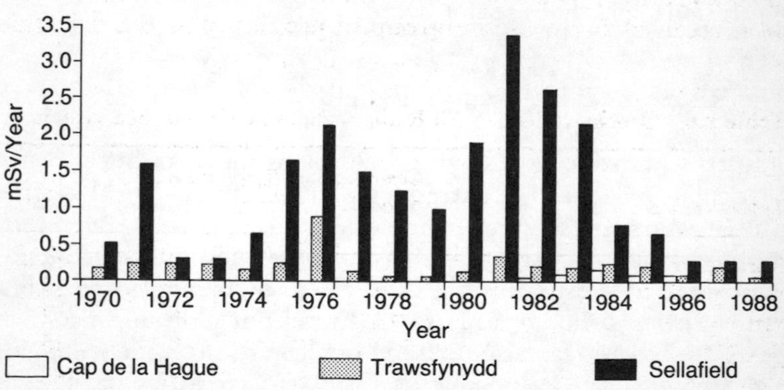

Figure 9.14 Highest Doses to Critical Groups at Three Sites (mSv/year). *Note:* Cap de la Hague Figures only available for 1982–86 for comparison.

absorbed. However, later evidence suggested this figure was too low and ICRP, reviewing these data in Publication 48 (1986), suggested an enhanced gut uptake figure of 10^{-3} (10^{-2} in the first year of life). These figures are about ten times the original value (a hundred times in the case of very young children). These values are acknowledged to be cautious as there is very little information to confirm them in real foods. Later work by MAFF using Cumbrian winkles fed to adult volunteers suggests that for americium, a gut transfer factor of 10^{-4} might be supportable but a factor of 2×10^{-4} was more reasonable for plutonium. Overall they suggest that a figure of 5×10^{-4} for Sellafield shellfish might be conservative. This figure, which means that a two-thousandths, or 0.05 per cent, of actinides are absorbed from shellfish, has been used in MAFF's 1988 report to estimate doses. It must be remembered that strictly it applies to shellfish and adults only, but as shellfish eaters are an important group in the Sellafield dose assessment context, use of this figure in the past would have reduced the critical group doses by factors of up to two.

Another rather worrying aspect of these assessments is the occasionally rather wide discrepancy between different doses calculated for what are manifestly the same populations. This is typified by the estimates made by BNFL and MAFF shown in Figure 9.15). Clearly the most conservative assumptions need to be employed in order to cover variations in eating habits, concentrations, etc. that are not overt.

The Risks of Living Near Sellafield

The critical group doses since 1970 paint an alarming picture, particularly if there are extreme members of the critical groups who are not represented (or found in habit surveys). Given that the discharges are justified and necessary, which will be disputed, the people of West Cumbria do seem to be having to receive a rather disproportionate amount of the risk. In addition, of course, people in other areas also polluted by the effluent from Sellafield, e.g. Ireland and the northwest European countries receive no balancing benefit from these operations.

The doses shown in Figure 9.14 have quite often breached the target dose (one-tenth of the public dose limit) which has been

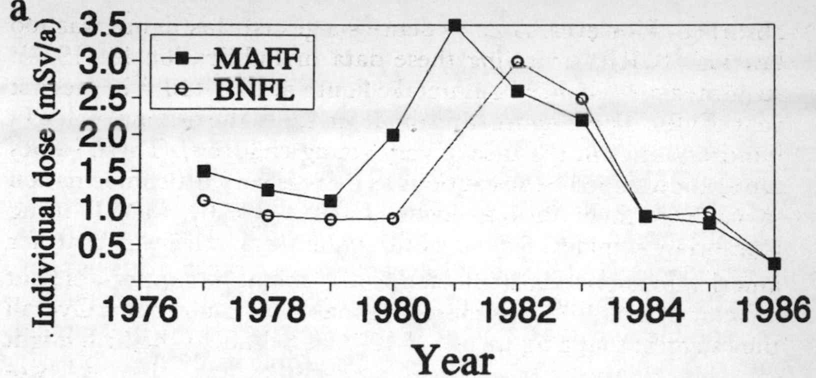

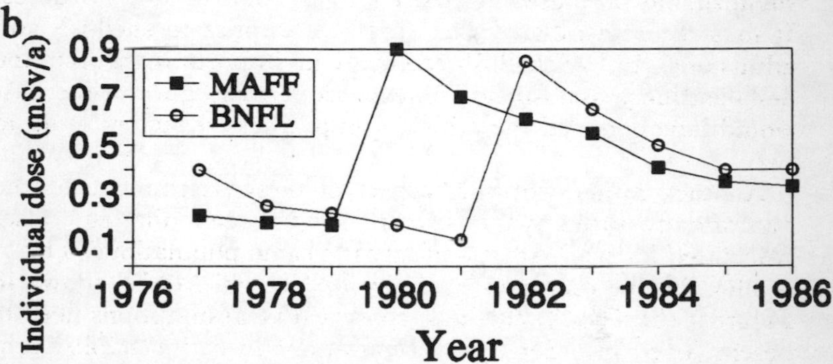

Figure 9.15 Critical Group Doses at Sellafield: (a) From Consumption of Seafood; (b) From Whole-body External Exposure.

used as a buffer. As a fraction of the then current public dose limit, the doses have remained fairly constant over the past 5–10 years, with an alarming peak in 1981. The annual total risk attendant on the average doses received over, say, the past five years is (using current risk estimates) about 1 in 32,000. This level of risk can be judged in a number of ways. The nuclear industry would compare the dose (and risk) with that received from natural background (see Chapter 6) and weigh the benefits of nuclear power. The environmental pressure groups would no doubt consider it an underestimate and point to the almost zero discharges achieved by other plants. However, as these are risks calculated from 'official' data, it may be instructive to compare this risk with what is considered tolerable by governmental sources. The Health and Safety Executive produced a report in 1987 (The Tolerability of Risk from Nuclear Power Stations) that stated that the maximum risk tolerable by members of the public from any large-scale industrial hazard would be 1 in 10,000. Risks of the magnitude of 1 in 32,000 would be in the region of 'just tolerable'. Later in 1987 the NRPB reviewed both the HSE report and a similar one by the Royal Society published in 1983 and came to the conclusion that a risk to the public of 1 in 100,000 should be considered just acceptable and that at 1 in 10,000 the risk would be definitely unacceptable. Thus the risks to some people in West Cumbria on the basis of 'official' or 'accepted' criteria are only just acceptable now.

Conclusions

All industrial processes, most of which in some way directly benefit society, produce waste. The production of energy is no different and it is abundantly obvious that coal extraction and burning produce waste that damages the environment and threatens life through air pollution. Somehow society accepts this as the price to pay and there is glib talk of 300 years of coal still to be mined and burnt. Energy produced by nuclear means is seen somewhat differently, with the industry constantly complaining about more rigorous standards. Unfortunately, because of early unfulfilled promises ('clean' power, 'too cheap to meter', etc.) and secrecy bred from the intimate connection with weapons production, the industry has a bad image. The public is not unnaturally afraid of radiation, a fact which the industry seems to have taken a long time to comprehend. Because of this anxiety, which is never going to be assuaged by bland reassurances or patronization, the industry must expect to have to work to high standards. The lack of a definite long-term policy to deal with intermediate and low-level waste is therefore, after 30 years of nuclear power, entirely reprehensible. Discharging waste into the sea was always a bad solution to a difficult problem and has been shown to be so. Although the new effluent treatment plants at Sellafield have dramatically reduced discharges, the legacy of only 10–15 years ago still pollutes the west Cumbrian environment and will always be there as a reminder of previous excesses. What is worse, the discharge authorizations seem to the uninitiated to have been driven by what the plant could manage rather than with the risk to the public in mind. If this is really the best we can do in environmental protection, something is surely wrong. Above all, looking at a map of concentrations of man-made radionuclides in the coastal waters of Europe, one is almost ashamed to be British. It is appreciated that the hazard to people in other European countries is small but it is not zero either. Do we have to continue polluting the seas if alternative technology exists?

How is the situation to be put right? There is nothing that can be done to reduce the contamination of the Cumbrian coastline. In fact, remobilization of silts and the decay of Pu-241 to Am-241 may well increase the risk from time to time. Sellafield's discharges have been reduced and the Enhanced Actinide Removal Plant will reduce them even further in the early 1990s, but is this enough? The throughput of THORP and decommissioning of Magnox stations is going to put a heavy strain on effluent processing. It certainly may breach the 'near zero release' ideal. The only really effective solution is to stop reprocessing nuclear fuel. Is this possible? Chapter 10 discusses this matter further.

10 How Safe Is Safe?

This book has been concerned with the effects of radiation on people and controversies that have mainly concerned the perception of these effects. The 'pollution' of the environment has therefore featured quite strongly in most chapters. How we perceive this pollution and whether we accept it as the price to pay for such things as employment and cheaper electricity are the contentious issues. 'Official' estimates of radiation risk or hazard often seem to be totally at odds with public perception and therefore discussion between the two parties is extremely difficult, especially as the concepts of radiation risk are not easily comprehended. In addition, government officials are only just learning the need to understand public anxiety.

One of the issues that has not been covered, but in the public's mind is misconstrued as being associated with the effects of radiation and radiation pollution of their environment is that of food irradiation. The irradiation of food for pasteurization purposes has not been covered here simply because the process does not make food radioactive and there are therefore no radiation effects on consumers to discuss. However, there is no doubt that this is a contentious issue for some of the public, in some part due to that misconception. Therefore a few words here may help to put the problem into perspective.

Food Irradiation

In 1990, it may become possible to buy irradiated food in shops in the UK. The government seems keen to lift the ban on food irradiation and confirm the recommendation made by its own Advisory Committee on Irradiated and Novel Foods about three years ago. Most consumers do not seem to want it and several of the big distributors have refused to sell it. However, more than thirty countries allow it for some foods and it may in the future be permitted throughout the EEC and so be difficult to resist.

The process preserves food by killing micro-organisms usually through exposure to a large gamma ray emitting source. The dose received by the food is recommended to be no more than 10,000 Gy. At this dose, the food is pasteurized (rather than sterilized) and will keep longer. Those who are in favour of the process point out that food is often already preserved by chemicals of dubious long-term toxicity and that an increase in shelf life without loss of quality could reduce distribution costs.

The rather formidable opposition has concentrated on the acknowledged loss of vitamin content, the introduction of radiolytic products of unknown long-term effect and the absence of a test that indicates whether or not irradiation has taken place. The last point is important; although radiation will kill bacteria, it will not remove the toxins they produced before they were killed. It is possible that unscrupulous people may sell food that has been 'cleaned up' by irradiation. In addition, of course, no matter how good the labelling, it will not be possible to know whether fast food or restaurant food has been irradiated. Clear labels are required, not 'pretty' ones (Figure 10.1), so that the consumer has a choice.

Figure 10.1 Symbol Indicating that a Food Product Has Been Treated with Ionizing Radiation.

There are drawbacks. Not all foods can be irradiated, and some taste a little different; not enough is known about the effects of irradiation on packaging materials, additives and pesticide residues; and some studies have shown that the irradiated food should be stored for at least a few weeks before being eaten. All in all, insufficient is known about the effects of irradiation on food for it to be allowed as yet and it certainly should never be introduced until a simple test is available to confirm it has been used on a food. Lastly, clear, unmistakable labelling is required.

The Reprocessing of Nuclear Fuel

One thread that runs through several chapters of this book is that nuclear power would be an altogether safer option under normal conditions if no fuel reprocessing was done. I would guess that similar thoughts must have occupied the minds of CEGB executives because nothing has done more to besmirch the name of nuclear-generated electricity than the discharges from Sellafield. Do we really need to reprocess?

It is worth examining the government's reaction to this question. This was given most recently in its response to the Environment Committee in 1986. In its Recommendation 19, the Environment Committee asked the government to prepare a detailed analysis of the financial consequences of abandoning THORP, and how to redeploy the THORP construction workers on 'cleaning up' the Sellafield site. The government rejected this recommendation and defended the construction of THORP and future reprocessing. This case is identical (as would be expected) to that set out by BNFL for the continuation of reprocessing Magnox fuel and for the construction of THORP. The main planks of the BNFL and government argument are:

(a) Dry stores are only a temporary option;
(b) BNFL has spent £500 million on replacement and refurbishment of the Magnox fuel-handling plant;
(c) Magnox fuel cans and uranium metal itself are susceptible to corrosion in water (in which they are now stored, except those at Wylfa power station);
(d) Dry storage/direct disposal would not reduce 'exposure to the general public or the nuclear industry workforce . . . it

would not have justified the extra cost';

(e) AGR fuel in stainless steel cans is also susceptible, after ten years, to slow corrosion in water. By 1992, when THORP should be operational, some AGR fuel will have been stored for fifteen years;

(f) Recovered uranium and plutonium can be fed back (recycled) to generate new fuel, particularly utilizing a mixed uranium/plutonium oxide fuel (MOX);

(g) The commercial success of THORP is already assured. Thirty utilities in nine countries have placed contracts worth £2.5 billion, and the CEGB/SSEB have placed contracts worth £1.6 billion;

(h) The abandonment of THORP, which has cost £1.6 billion to build, would lose this money, plus the overseas revenue. The employment consequences would also be severe in that 2000 construction workers are currently employed and 1000 BNFL employees will eventually staff the plant. The design, engineering and equipment supply presently support a further 1000 jobs in BNFL and 5000 elsewhere in British industry. Investment in THORP accounts for almost half of BNFL's current investment programme, which supports an estimated 50,000 jobs;

(i) The government remains firmly committed to THORP and reprocessing and there can be no question of abandoning the project.

Listed like this, the case looks overwhelming but does need some examination as it has such far-reaching environmental implications.

Reprocessing technology was originally developed in the 1950s to meet the military requirement for plutonium. This requirement is no longer as great and the fast reactor programme, using plutonium as a fuel, has also been halted so there is even less demand for plutonium. Generally, reprocessing increases the total volume of wastes by as much as 100 times (160 for Magnox fuel) although it can reduce the HLW by one-third.

It is clear from government and BNFL statements that about three-quarters of the throughput of THORP for the first ten years will be from foreign contracts. Quite apart from the added risk of returning plutonium and HLW, if any LLW or gases are to be dumped into the environment to expose the West Cumbrians, the only justification for this can be on the grounds of improving the balance of payments. It remains to be seen whether this is sufficient.

The case for reprocessing based on the need to prevent further corrosion of Magnox and AGR fuel cans is more reasonably based. However, the argument used in (d) above for not building dry stores for spent fuel seems on the face of it absurd, and needs detailed justification. Is it not possible (as has been suggested) to convert THORP for use as a dry store?

Britain and France are the only countries to reprocess fuel on any scale. However, France's operation is closely tied to the military need for plutonium for its independent nuclear weapons programme. West Germany has tried several times to build a commercial reprocessing plant; it has a small one at Karlsruhe and tried to establish a plant at Gorleben but this was abandoned in 1979. More recently, a plant at Wackersdorf was planned but has now had to be abandoned. Certain countries with significant nuclear power, e.g. Canada, have planned and implemented dry storage. Unless BNFL could aim for zero discharges, one feels the expenditure on THORP is misplaced.

Emergency Planning

Another controversial subject that has caused public anxiety is the apparent chaos in government circles after the Chernobyl incident. The media, on behalf of the public, were constantly asking how it was that after thirty years of nuclear power we had no early warning system or effective bureaucratic mechanism to arrange monitoring and disseminate information and advice. This subject was explored in some detail in Chapter 4 and the basic conclusion was that the chaos was caused by the lack of overall strategy and too many overlapping responsibilities in governmental departments. It does seem there is a need in the UK for an equivalent of the US Environmental Protection Agency (EPA) with wide-ranging powers.

It cannot be said that the situation has improved; the government continues to create committees with similar responsibilities, and the overall plan, which was first published in 1982 by the HSE as a booklet entitled 'Emergency Plans for Civil Nuclear Installations' is still awaiting update.

What has been done, however, is the setting up of the first phase of a national early warning system. Details of this system were first published by DoE (HMIP) in 1988 as a set of proposals that have since been partly implemented. These proposed a National Response Plan and a Radioactive Incident Monitoring Network (RIMNET). The Response Plan had been announced in June 1987 and was specifically to deal with nuclear accidents abroad. Again, the proposals are complex and involve a number of government departments but the RIMNET monitoring network does seem a step forward. There will be about eighty main monitoring sites throughout the UK (see Figure 10.2). These will be meteorological stations, defence establishments, government laboratories and NRPB centres. The sites will have remote recording gamma ray monitoring equipment that will eventually be connected to a computer in the DoE for interrogation and recording. Part of this network is already in existence and it is hoped the complete Central Database Facility (CDF) and Phase 2 will be operational by 1991.

This sort of equipment should have been available here many years ago as it was in countries such as West Germany and Sweden. It cannot prevent accidents, but by providing early warning of an incident, radiation doses could be reduced by, for instance, distribution of iodate (for thyroid blocking), sheltering or evacuation.

These proposals state that RIMNET 'will not be suitable for assessing the short range hazards of a nuclear installation accident in this country'. Procedures for such accidents, it states, 'have been in existence for many years'. This is a worrying statement. Indeed each nuclear facility has its own emergency plan with little responsibility for off-site monitoring or evacuation. However, there seems to be no central plan nor, in many areas, a County Emergency Plan. Co-ordinated evacuation would be a very difficult problem in this country and requires a central emergency plan adapted to the special problems of widespread radioactive contamination.

A Recapitulation

The general thesis of this book has been 'How safe is safe?' i.e. can we believe 'official' reassurances that the risks associated with

Figure 10.2 Initial Proposals for RIMNET Monitoring Sites in the UK.

exposure to radiation are 'negligible', 'insignificant', 'small compared to other risks', or 'no cause for concern'. It is hoped that the main conclusion the reader may derive is that there is such a dearth of evidence and so much uncertainty about environmental predictive models and radiation effects at low levels that it behoves the administration and regulation makers to be conservative. The public must look askance at governments who seem set on a policy of 'foot shooting'. Lack of forethought and underestimation or ignorance of public opinion lead to the loss of credibility very quickly.

In order to make these conclusions it is worth reviewing each chapter and emphasizing the important facets. Each chapter has been written with the aim of discussing a specific controversy; these are highlighted below.

Chapter 1 is informative but also covers the uncertainty that exists in our knowledge of the mechanism of radiation effects. There are several theories of carcinogenesis but still not enough is known, for instance, to enable us to extrapolate experiments on malignantly transformed cells in vitro to man's experience or to understand what happens in the latent period. There are very few data on effects on the immune system, and yet this would be a good candidate to explain some of the effects seen. The data on hormetic effects for radiation are not strong and are full of inconsistencies. Again, however, a threshold theory of cancer induction (biologically unlikely) would fit some experimental results. It must be said that the hormesis data are slightly less convincing than the reported effects actually seen at low doses.

Chapter 2 examined the sources of our information on radiation effects. If nothing else it emphasizes that by far the largest amount comes from experiences at high doses and at high dose rates. All the data have some drawbacks and there are proponents of each epidemiological study to be found who reject the others. In this minefield of information the assessor of risks needs once again to tread warily. What is in fashion now as explanation may not be so in ten years time as a population ages. This is particularly true of some of the medical studies for which the follow-up is fairly meagre. Clearly emphasis should be given to epidemiological studies of workers or members of the public (especially those in enhanced background areas) because these people have actually received the level of dose that is of interest. However, deductions drawn from studies with follow-up periods of less than twenty-five years can scarcely be conclusive. At the moment, of course, the complete dose record is often not known so that the risk part of the equation is a bit hazy.

As Chapter 3 indicated the actual estimates of risk and the setting of dose limits is even more uncertain. Too often estimates of risk are bandied around without any uncertainties attached and with no reference to the biology implied. The bodies recommending dose limits obviously need to work with a simple system of risk but the dependence on age at exposure, sex, etc. cannot continue to be ignored.

The introduction of factors to reduce risks that are becoming uncomfortably large has exposed inconsistencies. The introduction of a dose rate effectiveness factor is reasonable, biologically, but using a single figure for all cancers unless it is the most conservative (protective) is indefensible. Very few new data on the magnitude of this factor have been acquired in the last ten years and yet oddly it is perceived as more important and higher now. The system of dose limitation based on these shaky foundations of risk assessment needs to be restrictive and internally consistent, e.g. the use of risk weighting factors to calculate effective dose needs revision.

Another point on risk evaluation is a plea for less strict adherence to mathematical models. A case in point is the fitting of curves to the fatal cancer data from Japan. The idea behind curve fitting is the assumption that some function describes dose response. We do not really know what this function is, but the data seem to fit linear or linear-quadratic curves (as described in Chapter 3 a linear-quadratic is a dose–response relationship with dose + dose^2 function). This is an appropriate procedure although having fitted the curves there is a tendency to forget that they were not chosen because of a theory or knowledge of the underlying biological processes. There may be a tendency to imagine that the functions chosen represent a radiobiological truth. It must be remembered that it is an act of faith in mathematics to rely on such curves being valid well outside the range of the data points. Theories of radiation carcinogenesis should dictate which curves are fitted. A linear-quadratic suggests a two-hit model, which is a more plausible explanation, although a linear model is not excluded. Unfortunately much risk assessment at this time seems to be an exercise in curve fitting with slavish adherence to computational simplicity without even a passing nod to radiobiological theories.

There is obviously a need for some rethinking of the overall system of dose limitation, particularly as estimates of total risk rates are between five and ten times those used by ICRP at present. It also seems unlikely that lifetime fatal radiogenic cancer risk rates could be higher than about 10–12 per 100 per sievert, otherwise such effects would already be evident in the worker epidemiological studies carried out so far.

Lastly, in Chapter 3, the tolerability of risk was explored. Assuming a linear dose–response relationship also implies that at some level, radiation doses, and therefore risks, are considered 'tolerable' or 'acceptable'. This is, of course, in addition to the risk from natural background, about which we can do very little.

For the worker the acceptance of risk is deemed to be offset by being paid to accept it. However, workers in other jobs are also paid to accept risks, and uniformity among workers is a first step towards improving working conditions. Clearly, the best conditions should be sought.

In assessing tolerability of risk a common approach has been to make comparisons between fatal cancer risks and occupational accident rates. However this is not comparing like with like. The ICRP have sought to introduce an Index of Harm by quantifying all causes of death and expressing the resultant risk as 'days of life lost'. Using such methods, fatal accidents lead to a greater number of days lost than a fatal cancer because cancers normally occur later in life. It is doubtful whether a workforce would see the comparative risks in this way, but UNSCEAR in its latest report gives some assessments of life lost as a result of cancer using both the absolute (additive) and relative (multiplicative) risk models. The differences between the models expressed in this fashion is quite small (see Table 10.1).

A further comparison that has much to recommend it is that between fatal cancer rates from exposure to radiation and exposure at work to chemical carcinogens. The pattern of chronic exposure is similar but there are a couple of drawbacks; chemicals tend to be site specific and much of the information concerns exposure levels in the past compared to current levels, which are usually much improved. Consider comparison of exposure to asbestos (risk of mesothelioma and lung cancer) and benzene (risk of leukaemia) as examples, with exposure to radiation using the NRPB's risk rates.

Table 10.1 Summary of Projections of Lifetime risks for 1,000 Persons (500 Males and 500 Females) Exposed to 1 Gy of Organ Absorbed Dose of Low-LET Radiation at High Dose Rate. (Based on the population of Japan)

	Risk projection model	*Excess fatal cancers*	*Years of life lost*
Total population	Additive	40–50	950–1200
	Multiplicative	70–110	950–1400
Working population	Additive	40–60	880–1330
(aged 25–64 years)	Multiplicative	70–80	820– 970
Adult population	Additive	50	840
(over 25 years)	Multiplicative	60	620

Lifetime (35 years) risk of death from exposure to asbestos (chrysotile) at the 'control limit' (0.5 fibres per mil) is 8.8–15.3 per 1000.

Lifetime (45 years) risk of death from exposure to benzene at the 'action level' (0.5 ppm) is 5 per 1000.

Lifetime (35–45 years) risk of death from exposure to radiation at 15 mSv/yr is 18–23 per 1000.

On this basis the conclusion must be that radiation workers are underprotected compared with other industries.

Chapter 4 explored the government or official reaction to widespread, low-level radioactive contamination such as occurred after the Windscale Fire in 1957 and the Chernobyl incident in 1986, the point being, how much have we learnt in thirty years' experience of nuclear power and how ready are we to face the next nuclear incident?

The inescapable conclusion from the chaos in 1986 is that we have learnt little. If food is contaminated again, the government will have considerable difficulty in persuading the public and our European neighbours to accept the new contamination standards. The extreme reluctance of the government to plan for an event of low probability of occurrence of large effect (if it occurred on the French or Belgium coast) is reprehensible. The fact that the nuclear power industry in the UK survived the Chernobyl incident points towards an amazing tolerance on the part of the general public. Now is the time for the government to plan for the unthinkable.

Chapter 5 pointed out that the reality of childhood leukaemia clusters in statistical terms has been accepted by government committees (COMARE) and independent researchers. The general consensus is that, although the doses that can be calculated are insufficient with present knowledge to account for the leukaemias, the increased incidence does appear to be associated with the mere presence of nuclear plants. In this respect, the common factor seems to be a plant producing or using actinides, e.g. plutonium. It is to be regretted that an emotive topic such as this has been subject of such irresponsible reporting by the media. 'Pockets' of excess cancer are appearing everywhere and being reported without any regard for the statistics of epidemiology. However, there is research (funded, to their credit, by CEGB/BNFL) that attempts to identify possible routes of exposure of the foetus or new born that could account for an increased risk of leukaemia. Until this work is completed, any solution is pure speculation.

Chapter 6 on natural background radiation has been included mainly to illustrate the moral dilemma over increased levels of radon in homes and the official stance in terms of remedial action. Clearly, if a linear dose–response relationship is accepted, natural background should be responsible for up to 10 per cent of all cancers (more in children). Under these circumstances, a plea for financial assistance from householders whose dose exceeds, say, 15 mSv per year seems reasonable. This could surely be regarded as a house improvement grant.

In Chapter 7 it was shown that doses received from medical radiation could easily be reduced without losing diagnostic efficiency. It is odd that measures to do this have not been implemented to any great extent. There is no doubt of the benefits of X-rays, both for medical and dental purposes, but if 50 per cent of hospitals still do not have rare-earth screens, one wonders whether this is deliberate medical policy. At a time when the NRPB's surveys show the *average* dose from medical radiography increasing by about 20 per cent since 1983, it is surprising that such cost-effective measures are not insisted upon.

Chapter 8 examined all forms of transport and attempted to dispel fears about rail transportation. The public perception of the transport of radioactive materials is the sight of huge white, spent nuclear fuel flasks on rail trucks. Local authorities preach unreasonable doom and gloom over these flasks and declare their areas 'nuclear free zones'. Data presented in the chapter indicated that much of this concern is misconceived. The possibility of an increase in plutonium and HLW shipments to and from Japan is, however, not viewed with similar equanimity.

Chapter 9 on discharges to the environment, is one of the most important in that it illustrates the problems of unfounded confidence. As much as the government or BNFL would like to play down the presence of fission products and actinides in the environment, the contamination was not expected to be the problem it is. It demonstrates how little we know and how prudence must govern discharge policy in the future.

In the event, BNFL has invested enormously in plant to cleanse its effluents. Creditable as this is, however, it is too late. The legacy of the past will no doubt dog their attempts to reduce doses to critical groups much further. These doses, as was shown in Chapter 9, can be perceived as only just tolerable. What emerges loud and clear from these data is the need both for a defined policy for handling all types of radioactive waste and uniformity in target doses to critical groups whether they are exposed to a waste facility or an operational nuclear plant.

Overall I hope the message from these ten chapters is a call for a more scientifically reasoned attitude towards radiation risks. Both sides in the arguments have lost credibility; the officials by constant changes of policy or by simply having no policy and the pressure groups by making exaggerated and scientifically baseless claims. Clearly there must be some level of 'acceptable' or 'tolerable' risk from radiation. We live in a society full of risks, and radiation risks need management in the same way as other pollutants. Obviously constant vigilance is necessary and there should be constant pressure. But if we were to think of, for instance, abandoning nuclear power we would need to have a viable, pollution-free alternative (*not* burning more coal).

In the long run a more responsible attitude by government and the public to energy conservation would reduce demand and maybe encourage the search for cleaner alternatives. I only hope that the French and Belgians with their battery of nuclear power stations 40 km away across the Channel feel the same.

Glossary

Advanced Gas-Cooled Reactor (AGR) The AGR is a carbon dioxide cooled, enriched uranium fuelled, graphite moderated design that is unique to the UK. It operates at a higher temperature than the Magnox reactor (q.v.) and is more thermally efficient. In addition, because the fuel is clad in stainless steel, there are no great problems from corrosion. These reactors were developed from the Windscale AGR (WAGR). The first station was ordered by the Central Electricity Generating Board (CEGB – q.v.) in 1965 and was eventually completed at Dungeness in 1982, many years behind schedule. The CEGB eventually built seven of these nuclear power stations but most were years late in construction, expensive to build and operate, and usually near the bottom of the world reactor league table in availability (and therefore power production). In 1973, Arthur Hawkins, the then Chairman of the CEGB, told a government select committee that the AGR programme was 'a catastrophe we must not repeat'.

Alpha Particles Alpha particles are emitted by some unstable radioactive isotopes with a mass greater than 208 and also by a few lighter nuclides. Their energy spectum is of characteristic single peaks (usually between 4 and 6 MeV). They are helium nuclei and therefore doubly positively charged. Being big, heavy and charged, they are easily stopped (absorbed) by materials they encounter. They travel therefore only about 4 cm in air and a few tens of microns in biological tissue. They are stopped by a thin sheet of paper or the outer layer (epidermis) of the skin. Emission of an alpha particle reduces the atomic weight by 4 and the atomic number by 2, thus:

$$^{226}_{88}\text{Ra} \xrightarrow{\alpha} {}^{222}_{86}\text{Ru}$$

Alpha particles have a quality factor of 20. In other words, they are assumed to be 20 times as biologically damaging as gamma and beta radiation dose for dose.

Atomic Number The number of protons in the nucleus of an atom. This number represents the order of the element in the Periodic Table, and therefore determines its chemical properties.

Atomic Weight The weight of an element relative to C-12 as 12.000.

Beta Particles Beta particles are positively or negatively charged particles. They are emitted from many unstable radioisotopes when an intra-nuclear event occurs, changing either a neutron into a proton and a negatively charged electron, or a proton into a neutron and a positively charged positron. They interact with material they encounter, producing ionization. However, because of their light mass and high speed, some have considerable penetrating ability although their path may be tortuous. Their penetration may be up to several centimetres of tissue depending on their energy. They are emitted with a spectrum of energies up to a maximum (E_{max}) that is characteristic of the isotope. Some isotopes emit beta particles with only very weak energies, e.g. tritium (H-3) has an E_{max} of 0.018 MeV, C-14 has an E_{max} of 0.15 MeV but some are very energetic, e.g. P-32 has an E_{max} of 1.7 MeV and Y-90 (daughter of Sr-90) has an E_{max} of 2.2 MeV. Beta particles are effectively shielded by materials of low atomic number (q.v.), e.g. plastics such as Perspex, but when absorbed produce weak gamma radiation called Bremsstrahlung (literally 'braking radiation'). When a beta particle is emitted, the mass number is unchanged and the atomic number is increased by 1. For example:

$$^{90}_{38}\text{St} \xrightarrow{\beta} {}^{90}_{39}\text{Y}$$

Beta particle emitters do not generally present a serious hazard outside the body.

British Nuclear Fuels plc (BNFL) Formed in 1971 out of the former Production Group of the United Kingdom Atomic Energy Authority (UKAEA – q.v.) it became a public limited company in 1984. All the shares are held by the government. BNFL employs 15,000 people at plants concerned with nuclear fuel fabrication (Springfields), fuel enrichment (Capenhurst), reprocessing and high-level waste storage (Sellafield). BNFL also operate a reactor at Chapelcross that produces plutonium and electricity for the grid. It owns and controls a low-level waste dump site at Drigg south of Sellafield and its head office, and engineering and design departments are at Risley, near Warrington.

Cancer Otherwise known as a malignancy or neoplasm, it is a tumour or mass of cells out of control of the body's growth-limiting system. Cancers can occur in most tissues but they have different names depending on their origin; thus:

(a) carcinomas occur in epithelial cells, that is, the cellular covering of internal or external body surfaces, e.g. bronchogenic carcinoma is a lung tumour;

(b) sarcomas are cancers of connective tissue, e.g. osteogenic sarcoma is a bone tumour.

A number of physical and chemical agents are known to cause cancer, including radiation. There is always a latent or lag period before the cancer appears. Cancers caused by radiation are indistinguishable from cancers caused by any other agent. About one in four of the population dies of cancer, the incidence of which increases quite markedly after the age of about 65. Radiation-induced cancer is termed a stochastic effect, that is, the effect is governed by the laws of probability.

CEDE (Committed Effective Dose Equivalent) The effective dose equivalent weighted for organ risk summed over a fifty-year period, from an intake of radioactivity. It is in other words, the dose to which one is committed as a result of an intake. It is conventional to allot the dose to the year of intake.

Central Electricity Generating Board (CEGB) Shortly (1989) to become privatised, the CEGB was created by the Electricity Act of 1957 to be responsible for electricity generation and supply in England and Wales. The South of Scotland Electricity Board (SSEB) fulfils the same function for Scotland. The CEGB's network power is one of the largest centrally controlled systems in the world. Its nuclear capacity, at present about 17–20 per cent of the total, is derived from eleven ageing Magnox (q.v.) stations and seven advanced gas-cooled reactor (q.v.) stations. The Board is presently committed to building at least four pressurized water reactors (q.v.).

Chromosomes Structures within cells of living organisms that contain the deoxyribonucleic acid (DNA – q.v.). In all organisms except bacteria, viruses and the blue-green algae (prokaryotes), the genetic material is divided into a number of chromosomes, each carrying a unique subset of genes. When one cell gives rise to two daughter cells, each daughter receives a complete collection, in the process called mitosis. It is known from radiobiological experiments that cells in tissue culture are most vulnerable to the effects of ionizing radiation during mitosis. Gross chromosomal damage (chromosome aberrations) is easily observable in both plant and animal cells that have received certain doses of radiation. These sorts of changes probably result in cell death. More subtle changes in the infrastructure of the chromosomes, leading to mutations (q.v. deoxyribonucleic acid), may give rise to cancers if they occur in non-reproductive (somatic) cells, abnormalities or death in embryonic tissue, or to malformed or diseased offspring if it occurs in the reproductive cells of the parents.

Collective Dose The sum of all the doses received by a population and expressed as man-sieverts (man-Sv). It provides a measure of the collective risk to a population from a waste disposal practice, or from an accident, or other exposure.

Cooling Pond A large tank or pond in which spent (used) reactor fuel is placed to cool, both thermally and radioactively. Nuclear power stations have cooling ponds into which the fuel rods are placed immediately after removal from the reactor. The rods then stay in these ponds for periods of at least six months in order for short-lived fission products to decay. A typical pond at a Magnox station is 6–7 m deep and is cooled, filtered (and purified by ion exchange resins) and kept alkaline to minimize corrosion of the fuel cans. AGR cooling ponds contain boric acid, a neutron absorber, and are slightly deeper than those for Magnox fuel. There are also cooling ponds at the reprocessing plant at Sellafield (q.v.) into which the spent fuel rods are stored to await reprocessing. Although water in these ponds offers a cheap, simple means of shielding, it also has and continues to cause problems. Magnox fuel cans (made of magnesium alloy) corrode after long periods (one to two years) in water and allow fission products to escape. This has resulted in severe contamination of the cooling ponds at Sellafield and to large discharges of fission product waste into the Irish Sea in the 1970s. It also resulted in the construction of SIXEP (the Site Ion Exchange Plant) for 'cleaning' of the pond water. AGR fuel cans are made of stainless steel and PWR cans of a zirconium alloy that do not corrode appreciably in water. There are large quantities of spent fuel (AGR and foreign PWR) stored under water at Sellafield awaiting the completion of the oxide fuel reprocessing plant (THORP). The corrosion of Magnox fuel cans in water storage is one of the main reasons given for the continued need for reprocessing.

Deoxyribonucleic Acid (DNA) Complex molecules that contain all the information (genetic material) a cell requires to function and allow the cell to reproduce itself. Corruption of this information (mutations) by chemicals or radiation, called mutagens, may be repairable by the cell. Some damage is non-repairable, however, or the repair may be imperfect, and either cell death or a change in cell function ensues. The latter may manifest itself in a cancer. Although not proven beyond all doubt, it seems that the most likely target molecule for radiation injury in cells is the DNA.

Department of Energy (DEN) Under the Atomic Energy Act (1946), the DEN has a duty to 'promote and control the development of nuclear energy'. The Secretary of State for Energy appoints the board of the CEGB and also two representatives to the board of NIREX.

Department of the Environment (DoE) The DoE has formal responsibility for control of all forms of pollution. It issues authorizations jointly with the Ministry of Agriculture, Fisheries and Food (MAFF – q.v.) for disposal of radioactive waste into the environment. The DoE does some direct monitoring of water supplies and discharges. The main department of the DoE responsible for carrying out the provisions of the Radioactive Substances Act (1960) is Her Majesty's Inspectorate of Pollution (HMIP) which replaced the Radiochemical Inspectorate. The DoE also funds research into different aspects of radioactive pollution.

Ecology The study of the complex interrelationships between living organisms and their physical and chemical environments. Radioecology is the study of the behaviour of both natural and man-made radioisotopes in the environment.

Electromagnetic Radiation A waveform radiation continuum that varies in wavelength from 1500 m to 10^{-12}m, a range that encompasses radio waves, infrared, visible and ultraviolet light to X-rays and gamma rays. The wavelength of the waves is inversely proportional to the frequency, i.e.

$$\text{wavelength} = \frac{c}{\text{frequency}},$$ where c is the velocity of light.

Electrons Electrons are very small negatively charged subatomic particles that exactly balance the charge of protons in the nucleus. They are arranged in energy shells (or levels) around the nucleus, their configuration determines the chemical characteristics of the element. It is rearrangements in these shells, e.g. capture of an electron by the nucleus that can result in gamma emission from the atom.

Electronvolt A measure of the energy possessed by radiation. It is strictly defined as the energy imparted to an electron when it passes through an electrical potential of 1 volt. It is, however, applied to all forms of ionizing radiation. In the case of particles, their energy is dependent on their mass and velocity – hence, for an alpha particle and an electron to possess the same energy, the electron must travel at a very much higher velocity. The energy of electromagnetic radiation is inversely dependent on its wavelength, that is, short wavelength radiation contains more energy than that of longer wavelengths. The removal of an electron from an atom (the process of ionization) requires the input of energy of at least 13 electronvolts (eV). Ionizing radiation is generally very energetic, and this unit of energy is frequently expressed in multiples of a thousand (kilo electronvolts, keV) or millions of electronvolts (mega electronvolts, MeV).

Epidemiology The science of mapping of disease patterns in relation to parameters such as geography, time and socio-economic factors.

Equilibrium (Radioactive) This occurs between a parent and daughter radionuclide when the daughter decays as fast as it is produced from the parent. The daughter then appears to be decaying with the half life of the parent. Equilibrium can be set up between a series of short-lived daughters in a decay series, e.g. the short-lived daughters of radon.

Fast Breeder Reactor (FBR) A reactor that uses plutonium (P_u^{239}) as its fuel, and natural uranium (U-238) as a means of producing more P_u^{239}. As P_u^{239} is fissioned by fast neutrons, no moderator is required. Fast neutrons are also captured by U-238, which becomes U-239, which in turn decays to give P_u^{239}. Thus the reactor 'breeds' its own fuel. The reactor is cooled with liquid metal, e.g. liquid sodium. In thermal reactors, spent fuel contains both unused U-238 and P_u^{239}. Thus reprocessing of this material generates fuel for the FBR. A prototype FBR has been in operation at the UKAEA's site at Dounreay since 1974, and recently plans for an FBR reprocessing plant have been approved for construction at the site. However, a number of technical difficulties have been encountered in the operation of the FBR programme, and it is now under threat from lack of government support. It remains to be seen whether this will halt the building of the reprocessing plant, or whether the latter will simply be an extension to the same facilities in West Cumbria, a matter of concern to local environmentalists.

Free Radicals Molecular fragments (or sometimes atoms) containing an unpaired electron in the outer shell, as a consequence of which they are highly reactive, either due to their electron donating (reducing) properties, or their electron abstraction (oxidizing) properties. Ionizing radiation may inflict damage on biological material either by acting directly on 'target' molecules or by an indirect action, in which energy is deposited in water molecules to give rise to free radical species in water, the most significant being the highly oxidizing hydroxyl radical, $OH^{\cdot}$. These free radicals may then diffuse short distances to impart their free radical status on other molecules, forming organic free radicals. These in turn may eventually revert to their original state, or become altered. The chemistry of organic free radicals is complex but generally, when formed in vitro, oxygen adds rapidly to them to form what are usually even more reactive peroxy radicals. There is considerable scientific debate over the exact role of free radicals in the development of malignancies. However, there is good evidence that they may cause gene mutations, DNA damage and chromosome aberrations, and because any mutagen is potentially a carcinogen their role in such disease states is implicit.

Fission The phenomenon in which an atom of an element splits into two pieces almost equal in mass. This usually occurs because of bombardment by neutrons but elements of large atomic number undergo a significant fraction of spontaneous fissions as they decay. Fission was discovered in 1938 by Otto Hahn in uranium and is the process that produces power in a reactor (or in an atomic bomb). The power is developed by utilizing the fact that fission produces two new elements, the sum of whose atomic weights almost equals the fissioned element. The difference in mass (the mass deficit) is converted into energy that in a reactor manifests itself as heat. Another product of fission is neutrons. If these are of the right energy, they may produce further fissions in atoms nearby. Clearly if conditions are right, e.g. there is no absorption of neutrons, and neutrons of the correct energy are produced, the process can be repeated over and over again. This is known as a chain reaction and is encouraged in atomic bombs. The minimum mass of material in which a chain reaction is sustainable is known as the critical mass but 'critical' states may be controlled in far bigger masses. Power reactors are maintained at a just sub-critical state by control of neutron numbers (using control rods made of neutron absorbers like boron) and their energy (using moderators such as graphite to produce neutrons capable of causing further fission). The fission reaction may be damped down by accumulation of fission products that are neutron absorbers, e.g. Xenon-135; this is the reason for 'spent fuel' being removed from a reactor. Fission only occurs in elements of atomic number of 92 and above, and some of these are more easily fissile than others. For example, U-235 and Pu-239 are used as fuels in reactors. When natural uranium, which is predominantly U-238, is used in a reactor, the fission process occurs mostly to the scarcer U-235. In some reactors, the proportion of the more fissile U-235 has to be increased, a process called enrichment. The products of fission are generally intensely radioactive. The spectrum of elements produced shows two peaks at atomic masses of about 130–145 and 80–95. Thus typical fission products found in greatest abundance from the fissioning of uranium are Sr-90, Sr-89 and Cs-137, I-131, Ce-144, Ba-140, etc.

Fusion The process that occurs in the Sun to produce its heat and light. In certain conditions, isotopes of light elements, e.g. hydrogen, may be fused together to produce other elements. In this process, some mass is lost and this appears as energy. This is roughly the basis of the hydrogen (or thermonuclear) bomb and it is hoped that it will be the reaction taking place in a fusion reactor. In order to overcome the repulsive forces between the reactants, rather extreme conditions are necessary, as are found in the Sun, in particular a temperature of around 100 million degrees Celsius. Maintenance of these conditions for more than a few microseconds is proving very difficult. The fusion process used in the proposed reactors of the future will probably be:

$$^{2}_{1}\mathrm{H} = {}^{3}_{1}\mathrm{H} \rightarrow {}^{4}_{2}\mathrm{He} = \mathrm{n} = \text{energy (17.6 MeV)}$$

This reaction has enormous benefits to society because one of the reactants (deuterium H-2) is present in vast quantities in seawater, and the other (tritium H-3) can be generated within the reactor by the reaction of neutrons on a lithium blanket surrounding the vessel:

$$^{6}_{3}\mathrm{Hi} = \mathrm{n} \rightarrow {}^{3}_{1}\mathrm{H} = {}^{4}_{2}\mathrm{He} = 4.7\ \mathrm{MeV}$$

In terms of radioactive waste, there will be activation products from the effect of the neutrons on various components of the reactor, but generally these will not be as long-lived as fission products formed in the current type of reactors. Fusion research is very expensive and has benefited from international collaborative projects such as the JET (Joint European Torus) project at Culham and other European laboratories. Unfortunately there is no real evidence that commercial fusion reactors are 'just around the corner'; twenty-five to thirty years seems a more realistic timescale.

Gamma Radiation Part of the electromagnetic spectrum, this is one type of ionizing radiation with a wavelength of between 10^{-14} and 10^{-7} m. It is identical to X-rays, except for its origin. It is emitted by most radioisotopes during the course of their nuclear rearrangements, usually along with alpha or beta particles. It is highly penetrating, and lead or concrete are necessary to shield against its effects.

Half Life The fundamental law of radioactivity is that the rate of decay is proportional to the number of atoms present. This can be expressed mathematically as:

$$\frac{dN}{dt} = -\lambda N$$

where dN/dt is the rate of change of the number of atoms (N) with time (t) and λ is a constant (the decay constant). The integrated solution of this function is mathematically an exponential, that is the number of atoms will decrease by a constant fraction of those remaining in any constant time period e.g. by, say, 50 per cent in a year. In this case, the time of one year is called the half life (or half time) of that radioisotopic decay. It is simply the time taken for the radioactivity (measured in becquerels) to decrease by half. Half lives of radioisotopes vary from very small fractions of a second to many millions of years. Clearly, the specific activity (the number of becquerels per unit mass of the isotope) will vary with the half life, e.g. for an isotope with a very long half time, the number of, say, Bq per kg will be very low. This brings out the distinction between specific activity, which refers to Bq per gram of the isotope, and concentration, which refers to Bq per gram of sample material.

Ion An electrically charged atom, with either a positive or negative sign due to loss or gain of an electron respectively.

Isotopes Isotopes are different forms of the same chemical element. They differ only in the number of neutrons in the nucleus but have the same number of protons (the number characteristic of the element). Their chemical behaviour is thus typical of their element but they have different atomic weights. Some elements have several isotopes whereas some have only a few, e.g. hydrogen has three, namely protium (${}^{1}_{1}H$), deuterium (${}^{2}_{1}H$) and tritium (${}^{3}_{1}H$). If the imbalance between protons and neutrons in the nucleus is great the isotope will be unstable and will attempt to return to a stable state by emitting radiation. This is a radioisotope.

Leukaemia A cancer of the blood-forming organs. It is characterized by widespread proliferation of white blood cells (leukocytes) usually associated with the appearance of immature leukocytes in the blood. The classification of leukaemia depends on the clinical course of the disease and the type and maturity of the cells involved. Thus leukaemia may be chronic or acute and myeloid (from the bone marrow) or lymphatic (from the lymphatic system). Acute leukaemia can occur at any age, but in children it is usually lymphoblastic and peaks in incidence in the age range 3–6. In adults, acute leuka mia is more often myeloblastic. Chronic leukaemia is a disease of middle life (say 30–60) although chronic lymphatic leukaemia usually occurs at later ages. Treatment is generally more successful for the acute versions than the chronic. There are several related diseases that occur in the bone marrow and lymphatic system, such as the lymphomas, Hodgkin's disease and myeloma. The bone marrow is particularly sensitive to radiation and there is clear evidence of causation for the leukaemias except for chronic lymphatic leukaemia. There is no evidence as yet of a link between Hodgkin's disease (which may have a viral origin) and radiation.

Light Water Reactor (LWR) A reactor using ordinary water as a coolant and moderator; the term encompasses boiling and pressurized water reactors.

Linear Energy Transfer (LET) A measure of the distribution of energy along the path or track of an ionizing particle, or ray, as it passes within a material. LET is quoted in kilo electron volts (keV) per micron of track. LET is usually given as the average over the whole track length, thus smoothing out the peak (so-called Bragg peak) in ionization density near the end of some high LET tracks, e.g. alpha particles.

Magnox Reactors These power reactors are unique to the UK (only two have been sold abroad – the Latina station ordered by Italy in 1957 and the Tokai Mura ordered by Japan in 1959). The fuel, natural uranium canned in a magnesium alloy ('Magnox'), is located in a massive graphite block moderator. They are cooled by carbon dioxide gas and are relatively thermally inefficient. However, Britain has built eleven of these expensive power stations and they seem capable of exceeding their design life. The first of these nuclear power stations (Berkeley) has just been closed and a second (Bradwell) is scheduled to close within the next two years. These reactors were developed from the early air-cooled designs at Harwell and subsequently at Windscale (now Sellafield). The first commercial station, the Calder Hall reactor, opened in 1956 at Windscale and also produced plutonium for the weapons programme. The last was the Wylfa station which was ordered in 1963. Although some have been fairly reliable, the Magnox stations generally had problems in construction and the later ones had corrosion difficulties. All the Magnox reactors are slightly different in construction as design has evolved and they were built by a number of different consortia. This has compounded the difficulties of maintenance.

Mass Number The total number of particles in the nucleus of an atom, i.e. protons and neutrons. This is frequently given as a superscript before an element's symbol, while the atomic number appears as a subscript, for example:

$$^{235}_{92}U, \quad ^{137}_{55}Cs, \quad ^{3}_{1}H$$

The mass number is also often given as a number after the element's symbol, e.g. U-235, Cs-137, H-3.

Ministry of Agriculture, Fisheries and Food (MAFF) This ministry is responsible, inter alia, for the purity of food. In this role, it monitors both the aquatic and terrestrial food chains for radioactive contaminants and, jointly with the Department of the Environment (DoE – q.v.), sets limits on what and how much radioactive waste may be disposed of to the atmosphere and marine or freshwater environments. MAFF publishes comprehensive annual reports on discharges, monitoring results and critical group dose estimates.

National Radiological Protection Board (NRPB) A quasi-autonomous non-governmental organization, the NRPB was set up by the Radiological Protection Act (1970). It was formed by a merger of a Medical Research Council Unit, the Radiological Protection Service, and the Radiological Protection Division of the Health and Safety Branch of the United Kingdom Atomic Energy Authority (UKAEA – q.v.). Its remit was:

> by means of research and otherwise to advance the acquisition of knowledge about the protection of mankind from radiation hazards, and to provide information and advice to persons (including government departments) with responsibilities in the United Kingdom in relation to the protection from radiation hazards either of the community as a whole or of sections of the community.

The NRPB headquarters was located on the Harwell site just outside the fence of the Atomic Energy Research Establishment, and had three regional centres. Having been formed from part of the UKAEA, ties with the Authority were initially strong. Certainly the independence and impartiality of the NRPB were often questioned. A change of address from Harwell to Chilton did nothing to lessen this. However, more recently the views of the NRPB have become a little more radical. In 1987, it was joint host with Friends of the Earth of a conference on the Effects of Low-Level Radiation and, also in 1987, it upstaged the ICRP by suggesting a lowering of dose limits. Unfortunately they are still often seen as part of the 'Establishment'. The Board has twelve members of whom the chairman is at present Sir Richard Southwood. The Radiological Protection Act also provided the NRPB with an Advisory Committee of twenty-four members drawn from a wide cross-section of groups with interests in radiation protection. This might have been a useful consultative committee, but about eleven years ago, the government, in its wisdom, abolished it.

Nuclear Industry Waste Executive (NIREX) The government agreed in 1982 that the component parts of the nuclear industry should set up NIREX to enable them to fulfil their responsibilities for management of radioactive wastes that they generate. UK NIREX Ltd is a public company with ownership divided between BNFL, CEGB, UKAEA (qq.v.) and the South of Scotland Electricity Board (SSEB). The Secretary of State for Energy has a special share. It is NIREX's responsibility to provide, pay for, and manage new facilities for the disposal of low- and intermediate-level radioactive waste whether this is to be into a near surface facility on land, into a deep facility on land or under the sea, or onto the sea bed. At the moment, the second of these options appears to be the most seriously considered and sites under active investigation are the Sellafield area in Cumbria and the Dounreay area in Scotland.

Nuclear Installations Inspectorate (NII) A branch of the Health and Safety Executive (HSE) responsible for issuing site licences for nuclear installations. (The HSE is responsible for enforcing the Health and Safety at Work Act, 1974, and the Nuclear Installations Act, 1965; both concerned with safety in nuclear issues.) The NII's work is mainly concerned with the design and operational safety of the plant. It is responsible for ensuring that emergency plans exist and also the maintenance of records of all incidents at nuclear sites; the plant operator is required to keep a 'Register of Site Incidents' as directed by the NII.

Neutrons Uncharged sub-atomic particles of equal mass to protons present in the atomic nucleus. Within an element, different numbers of neutrons give rise to different isotopes. In most stable isotopes, the number of neutrons in the nucleus equals the number of protons. Where the ratio of neutrons to protons exceeds one, instability usually occurs, this effect increasing with increasing ratios. Such atoms undergo a nuclear rearrangement, resulting in radioactivity. In some high atomic number elements, bombardment with neutrons leads to fission and this in turn releases more neutrons. Neutrons may have different energies; those released as a result of fission are generally fast. While some isotopes are fissioned with fast neutrons, e.g. Pu-239, others require slow or thermal neutrons (e.g. U-235). Thus in an ordinary nuclear reactor, where the chain reaction to be sustained is that due to the fissioning of U-235, neutrons must be slowed down by a moderator. Neutrons may be captured by other atoms, in a process known as activation. This is a widely utilized process in analytical chemistry and in the production of isotopes for medicine. Unfortunately, however, it also leads to induced radioactivity in reactor materials.

Photon Gamma rays from radioisotopes are emitted in discrete 'packets' of characteristic energies. These packets are called photons.

Pressurized Water Reactor (PWR) This is a version of the light water reactor (LWR – q.v.). This type of reactor uses ordinary water both as a moderator and coolant, and enriched uranium (in which the more fissile isotope, U-235, has been increased in proportion to 2–3 per cent) as a fuel. The reactor is operated under pressure to prevent the water from boiling; in a variant, the boiling water reactor (BWR), the pressure is less and the water actually boils. Both these reactors have a relatively small, compact core and a high power density. The PWR was a direct offshoot of American efforts to produce a reactor suitable for submarine propulsion. The first US PWR was built at Shipping Port near Philadelphia in 1957. This reactor type is the commonest, being responsible for most nuclear generating capacity worldwide. PWRs are relatively cheap to build but rely on the integrity of the water cooling system because of their high power density. Hence a loss of coolant accident can quickly result in serious melting and distortion of the fuel assemblies; this has occurred on several occasions. Britain has decided to build four PWRs as the next generation of nuclear power reactors. The first is well under way on the Sizewell site in Suffolk, the second (subject to a planning public inquiry) is destined for the Hinkley Point site in Somerset. The radiation doses to PWR workers (particularly maintenance workers) are generally higher than for the Magnox/AGR plants. To meet their own dose targets, the CEGB will need to operate its new reactors to some of the best PWR standards in the world.

Quality Factor (QF) Strictly, a physical factor depending only on LET, i.e. on the density of ionization caused by the radiation. For example, an alpha particle produces about 1 million ion pairs per millimetre of track in tissue, whereas a beta particle produces only about 10,000. The QF of gamma rays and beta particles is 1, that of alpha particles, fast neutrons and protons 20 and thermal neutrons, 5. The QF is combined with absorbed dose to give dose equivalent (the sievert).

Radioactive Waste Management Advisory Committee (RWMAC) A committee set up in 1980 to advise the Department of the Environment on major issues relating to the development and implementation of an overall policy for the management of civil radioactive waste, including the waste management implications of nuclear policy, of the design of nuclear systems and of research and development, and the environmental aspects of the handling and treatment of wastes. There are nineteen members on the committee, drawn from academia, unions, CEGB, UKAEA and BNFL. It is a limitation on this committee that its remit is to consider only civil waste, particularly when a number of nuclear submarine reactors are to be decommissioned soon. It is also debatable how much notice the government takes of it; for exmple, RWMAC seem to be in favour of sea disposal for certain wastes whereas the government announced in 1988 that it had decided not to resume sea dumping. Although on this occasion the government had good reason to ignore RWMAC, it is ironic that this may well be the fate of parts of the reactors of old nuclear submarines.

Reactor A device for extracting energy from the fission process and using it (usually) to heat water to steam which then drives an electricity-generating turbine. Nuclear reactors come in different designs but are of two basic types, thermal reactors or fast breeder reactors, depending on their utilization of neutrons of different energies (thermal or fast) for the fissioning process. These are described under the appropriate headings elsewhere in the glossary. Nuclear reactors need fuel (normally uranium, either natural or enriched), a moderator to slow neutrons down to thermal or fissioning energies, and a coolant (to transfer the heat produced). When used on land, they are physically huge structures because of the need for metres of concrete shielding and containment.

Sellafield A BNFL site on the west coast of Cumbria that reprocesses nuclear fuel. It also has facilities for high-level waste storage (stainless steel tanks) and, soon, a vitrification plant. There is also a power reactor (Calder Hall) which was first to produce electricity in the UK. Sellafield is also the site of the two original air-cooled plutonium reactors (one of which was involved in the fire of 1957) which are now disused. The original AGR design (WAGR) reactor with its characteristic spherical containment is now being decommissioned. Sellafield was originally named Windscale; cynics believe the name change was to sever connections with the 1957 Windscale fire.

Tailings What is left behind when uranium has been removed from its ore. As the ore normally contains all the other radionuclides of the uranium decay series, including radium and its daughters, these are present in 'tailings'. The material therefore represents a significant hazard, albeit of natural origin, and has been responsible for enhanced pollution of rivers in the USA. It has also been irresponsibly used as landfill, most notoriously in Grand Junction, Ohio.

Ultraviolet Light (uv) Light that is just at the (blue) end of the visible light spectrum. The wavelength varies from about 10^{-7} to 4×10^{-7} metres, and is non-ionizing. However, it is absorbed specifically by certain molecules (chromophores), and can produce late effects at these sites, skin tumours being a particularly good example.

United Kingdom Atomic Energy Authority (UKAEA) The UKAEA was set up under the Atomic Energy Act 1954 out of the Division of Atomic Energy of the Ministry of Supply, which was primarily concerned with nuclear weapons production. It was responsible for the first UK reactor at Calder Hall and promoted the Magnox design. It is now carrying out research into a number of aspects of energy production and environmental protection. The main laboratories are at Harwell, Winfrith (reactor research) and Dounreay.

X-rays A form of penetrating electromagnetic radiation produced when electrons are accelerated into a metal target by a large voltage. The wavelength and therefore the penetrability of the X-rays depends on the accelerating voltage. Diagnostic X-rays are effectively shadow pictures that depend on X-rays' properties of being absorbed by bone and blackening photographic film. Diagnostic X-rays are generated by voltages of 50–150 kV whereas X-rays used for treating cancer are produced using voltages of several megavolts.

Further Reading

General

The books, papers and reviews listed here are those which are not only readily accessible but are also likely to be more easily understood. Two types of publication are recommended, those which present data or reviews of data and those which make comment on the data. The second group, unfortunately, can be further subdivided depending on the bias of the author. It will be apparent that a clear, unbiased, objective stance is a rare commodity in the radiation field.

For comprehensive reviews of data or environmental measurements, global doses and risks, the United Nations Scientific Committee on the Effects of Atomic Radiation (UNSCEAR) Reports are very useful (but also expensive). These reports have been published regularly for about thirty years but the 1982, 1986 and 1988 reports contain the most relevant data.

Another series of reports which are extremely valuable for both data and, to a certain extent, comment are the National Council on Radiological Protection and Measurements (NCRP) Reports from the USA. Many of the earlier reports are now out of print but the subjects are being constantly brought up to date, and there are over a hundred published now which represent a valuable data-base. However, they mostly concern radiation protection from the US viewpoint.

The NRPB now has a world-wide reputation for its reports which are an invaluable source of data, covering all aspects of radiation protection, dose and risk assessment, environmental modelling, etc. They are cheap and easily obtainable in the UK. Although they certainly do not present the last word on each topic, they are pragmatic and provide ammunition for both sides of the nuclear debate. Their report (R226) which explains the 1988 UNSCEAR report is required reading.

At this point it is worth recommending *Radiation and Health* (John wiley, 1987) which is a book of the proceedings of a conference organised jointly by Friends of the Earth and the NRPB at Hammersmith Hospital on 24–25 February 1986. The papers presented cover the whole gamut of environmental contamination and risk and the book is a unique contribution to the 'nuclear controversy' debate.

Lastly, in this general review, another book with the same title deserves a mention, namely, *Radiation and Health* written by Martin Dace for the Medical Campaign Against Nuclear Weapons (1987). This really is an example of 'everything you wanted to know but couldn't find anyone to ask', presented in simple language.

Recommended further reading for each chapter has again been kept to those documents, books, etc. which are readily available or can be obtained by request at a scientific library.

Chapter 1

There are a number of good books and booklets on the basics of radiation, its interactions with matter, instruments, units, etc. which might aid the reader. The NRPB produces a booklet, 'Living with Radiation' (now in its 4th edition, 1986, and available at HMSO), which will be found useful. The United Nations Environmental Programme's book *Radiation: Doses, Risks and Effects* (1985) has some excellent graphics (including Figure 1.2). For more advanced reading on radiation protection, Alan Martin and Samuel Harbison's book *An Introduction to Radiation Protection* Chapman & Hall, 3rd edition, 1986) provides comprehensive cover. One of the USA National Council on Radiation Protection and Measurements Reports ('A Handbook of Radioactivity Measurements Procedures' – NCRP No. 58), although rather dated (1985), is a good review of instrument usage theory.

There is a plethora of books and papers on all aspects of radiobiological theory. John Coggle's book, *The Biological Effects of Radiation* (Wyekham Publications Ltd, 2nd edition, 1983), is the most readable and should be required reading. In addition, Eric Hall's *Radiobiology for the Radiologist* (Harper & Row, 1978), in spite of its rather specific sounding title, provides good coverage of this subject for those with a biological background.

Chapters 2 and 3

These chapters are expressly about sources of information (data) on radiation risk. For further reading, the UNSCEAR reports are useful reviews as are some of the papers in *Radiation and Health* (John Wiley; recommended above). For both these chapters it is worth referring to two of the reports of the committee set up jointly by the US National Institutes of Health and National Academy of Sciences, the Biological Effects of Ionising Radiation Committee (so-called, BEIR Committee). The BEIR III report of 1980 and the BEIR V report of 1989 are useful (if complicated) reviews of radiation risk and somewhat different in approach from UNSCEAR.

Two recent books provide useful commentaries on the available data on radiation risk, these are: *Radiation Risks: An Evaluation* by David Sumner Tarragaon Press, 1987) and *The Dangers of Low Level Radiation* by Charles Sutcliffe (Avebury, 1987). A third book, *Multiple Exposures* by

Catherine Caufield (Secker & Warburg, 1989), is a more extreme and less scientific view.

It may also be worth referring to the proceedings of some of the recent conferences on the health effects of low doses of radiation. This subject is definitely 'the flavour of the month' as evidenced by the number of meetings and conferences recently (1989) organised to discuss it. Such conferences have generally included papers and discussion not only of conventional interpretations but also of more radical ideas, i.e. The British Nuclear Energy Society (1987) *Health Effects of Low Doses of Radiation – Recent Advances and Their Applications*; IBC (1989) *Effects of Small Doses of Radiation*; L. H. Gray Conference (1989) *Low Dose Radiation – Biological Basis of Risk Assessment* (Taylor & Francis).

For the most comprehensive review of current arguments on radiation risk, epidemiology, etc., the reader could well spend some time browsing through the transcripts of the evidence represented in 1989 to the public inquiry held to consider the CEGB application to build a PWR reactor at Hinkley Point in Somerset.

Chapter 4

There has been a vast number of papers and articles written on the effects of the three major nuclear accidents – some factual and some speculative – but far too many to list. The least covered accident is the second biggest, at Kyshtym in the southern Urals in Russia on 29 September 1957 (just about the time of the Windscale Fire), which only now (1989) has been reported in the Russian scientific press and to the IAEA.

The 'official' report of the Windscale Fire (Cmnd 302, HMSO), published in November 1957, makes very interesting reading when compared with the two NRPB reports (R-135 and Addendum) published in 1983.

As regards the Chernobyl accident, the NRPB published a *Preliminary Assessment of the Radiological Impact of the Chernobyl Reactor Accident on the Population of the European Community* which is comprehensive but already out of date in terms of risk projection.

The World Health Organisation (WHO) produced a useful 'Report of Consultation' (ICP/CEH 129) on the accident which lists the actions taken in various countries.

In the UK, all immediately available 'raw' monitoring data were published shortly after the accident (Department of Agriculture for Northern Ireland, DoE, etc., *Levels of Radioactivity in the UK from the Accident at Chernobyl, USSR, on 26 April 1968: A Complation of the Results of Environmental Measurements in the UK*. In addition, MAFF produced separate compilations of food monitoring data in 1987 and 1988, both entitled, *Radionuclide Levels in Food, Animals and Agricultural Produce. Post Chernobyl Monitoring in England and Wales*. All three publications are obtainable from HMSO.

Friends of the Earth produced a 'pack' of press cuttings and reports which can be read to judge the impact of Chernobyl as reported in the media. The UK government's actions are charted in 'Chernobyl: An Enquiry through Parliamentary Questions' by David Clarke, MP who was shadow Agriculture Minister at the time.

One of the most comprehensive accounts of the deposition of radioactivity in the UK has only just (1989) been published. This paper, 'The Transport and Deposition of Airborne Debris from the Chernobyl Nuclear Power Plant Accident with Special Emphasis on the Consequences to the United Kingdom' by F. B. Smith and M. J. Clark, Scientific Paper No. 42 from the Meteorological Office (obtainable from HMSO), belatedly demonstrates correlations between weather conditions, deposition patterns and uptake into dairy products.

Lastly, the Watt Committee produced a report, *The Chernobyl Accident and Its Implications for the UK* (Watt Committee Report No. 19), which is interesting (for instance, Table 4.6 comes from this report) but tends to play down the impact of the accident.

Chapter 5

Once again, the subject of 'leukaemia clusters' and epidemiology has received a great deal of attention, particularly since 1983. The columns of *The Lancet* and *Nature* have in the last few years often been the scene of fairly lengthy exchanges of letters on the subject.

The 'Report of the Inspector in the Windscale Planning Inquiry' and the 'Report of the Black Committee' (both obtainable from HMSO) are useful reading to set the scene.

The later reports (three so far) by COMARE (also from HMSO) provide some useful reading, as does the NRPB evidence to COMARE (R171 also read R170, R195, R196, R202) and the subject was covered fairly fully in *Radiation and Health* (John Wiley, 1981, see above).

Chapter 6

Much of the information about background radiation may be gleaned from the various UNSCEAR reports (particularly 1988) and the NCRP reports. The UNEP booklet referred to in Chapter 1 and the NRPB's 'Living with Radiation' are also useful guides.

The radon problem is also covered in NRPB reports (R152, R189, R190, R208, R227, R229) and in the BEIR IV report (The Report of the Biological Effects of Ionising Radiation Committee of the US NIH-NAS) published in 1989, *The Health Risks of Radon and Other Internally Deposited Alpha Emitters*.

Chapter 7

Further reading on the subject of the risks of medical radiation could be confined to UNSCEAR and NRPB reports (R104, R200, R201). The

NRPB especially looks upon the risks from medical radiation as an example of failure of the principle of optimisation in radiation protection. The ICRP (in publication nos 33, 44 and 52) and the NCRP with reports such as *Exposure of the US Population from Diagnostic Medical Radiation* (Report No. 100, 1989) and *Nuclear Medicine – Factors Influencing the Choice and Use of Radionuclides in Diagnosis and Therapy* (Report No. 70, 1982) have not ducked the issue. Apart from these, papers and articles by Dr J. G. B. Russell, who has conducted an almost one-man crusade in the UK to secure improvements in X-ray techniques and equipment in order to reduce patient risk, are worth reading.

Chapter 8

Extra reading on the regulations and hazards of the transport of radioactive materials could be limited to an NRPB report (R155) and to the reports of the Advisory Committee on the Safe Transport of Radioactive Materials (ACTRAM) (both available from HMSO).

Chapter 9

For those interested, the Radioactive Substances Act (1960) and amendments to it in the Environmental Protection Bill (1989) may be obtained from HMSO.

a comprehensive review of environmental monitoring was produced by HMIP in 1987 under the title *Monitoring of Radioactivity in the UK Environment*. The Watt Committee has produced a similar document for the nuclear industry entitled *radiation Monitoring Around CEGB Nuclear Power Stations*. As regards the actual monitoring results in terms of impact on the environment, MAFF (Annual Reports of Monitoring of the Aquatic Environment and of the Terrestrial Radioactivity Monitoring Programme), the Scottish Development Department (Environmental Monitoring for Scotland), and some site operators (CEGB, BNFL and UKAEA) publish monitoring reports, those by BNFL and UKAEA being annual.

Comment on discharges, environmental impact, etc. can be found in the publications of the Political Ecology Research Group (PERG). They produced a useful report (RR8) in 1982 entitled *Impact of Nuclear Waste Disposals to the Marine Environment*. This was followed by a paper by Peter Taylor published in *Radiation and Health* (John Wiley, 1987; referred to previously).

Policies for waste management are frequently discussed in ATOM (the journal of the UKAEA) and the *International Atomic Energy Agency Bulletin*.

The UK government's response to the Select Committee on the Environment (Cmnd 9852, 1987) makes very interesting further reading, as do the annual reports of RWMAC (from HMSO).

Naturally, both Greenpeace and Friends of the Earth have published a number of reports concerning waste discharges and monitoring. FoE are currently (1989) interested in tightening the regulations applied to non-nuclear establishments.

Peter Bunyard's book *Health Guide for the Nuclear Age* (Papermac, 1988) is full of useful data and makes some tellng points on environmental contamination.

There has also been a series of reports from the Atomic Energy Authority's Environmental and Medical Sciences Division at Harwell (obtainable from HMSO) on the behaviour of discharges from Sellafield in the environment (Figures 9.9 and 9.10 and Table 9.7 come from some of these reports).

For those who wish to read further on the science of the behaviour of radionuclides in the environment, R. J. Pentreath's book *Nuclear Power, Man and the Environment* (Taylor & Francis, 1980) is recommended as the most easily understandable.

Chapter 10

Food Irradiation: A number of informative (but 'anti') books on food irradiation has been written. suggested reading is *Food Irradiation: The Myth and the Reality* by Tony Webb and Tim Lang (The London Food Commission, 1989) and *The Biology of Food Irradiation* by David Murray (John Wiley, 1989). *Emergency Planning:* There is no easily available cheap book on this topic. There has been a number of symposia and conferences on the subject since Chernobyl and occasionally the proceedings have been published, the most recent being *Medical Response to Effects of Ionising Radiation*, edited by W. A. Crosbie and J. H. Gittus (Elsevier Applied Science, 1989). The subject has also been explored in some detail in the planning inquiries at both Sizewell and Hinkley Point. A critical review of emergency plans was presented to the Sizewell Inquiry by PERG under the title, *A Critical Review of Emergency Planning for Reactor Accidents and Spent Fuel Transport in the United Kingdom* (Research Report RR-12 by Roger Kayes and Peter Taylor).

In addition, emergency plans are availalbe to members of the public for each nuclear station in the UK.

Lastly, if you wish to balance your reading, the case for nuclear power is explored in: *Power Production – What Are the Risks* by J. H. Fremlin (Adam Holger, 1985) and *Before It's Too Late* by Bernard L. Cohen (Plenum, 1983).

Index